SCIENCE AND TECHNOLOGY POLICIES FOR THE ANTI-TERRORISM ERA

NATO Science Series

A series presenting the results of scientific meetings supported under the NATO Science Programme.

The series is published by IOS Press and Springer Science and Business Media in conjunction with the NATO Public Diplomacy Division.

Sub-Series

I. Life and Behavioural Sciences	IOS Press
II. Mathematics, Physics and Chemistry	Springer Science and Business Media
III. Computer and Systems Sciences	IOS Press
IV. Earth and Environmental Sciences	Springer Science and Business Media
V. Science and Technology Policy	IOS Press

The NATO Science Series continues the series of books published formerly as the NATO ASI Series.

The NATO Science Programme offers support for collaboration in civil science between scientists of countries of the Euro-Atlantic Partnership Council. The types of scientific meeting generally supported are "Advanced Study Institutes" and "Advanced Research Workshops", although other types of meeting are supported from time to time. The NATO Science Series collects together the results of these meetings. The meetings are co-organized by scientists from NATO countries and scientists from NATO's Partner countries – countries of the CIS and Central and Eastern Europe.

Advanced Study Institutes are high-level tutorial courses offering in-depth study of latest advances in a field.
Advanced Research Workshops are expert meetings aimed at critical assessment of a field, and identification of directions for future action.

As a consequence of the restructuring of the NATO Science Programme in 1999, the NATO Science Series has been re-organized and there are currently five sub-series as noted above. Please consult the following web sites for information on previous volumes published in the series, as well as details of earlier sub-series:

http://www.nato.int/science
http://www.springeronline.nl
http://www.iospress.nl
http://www.wtv-books.de/nato_pco.htm

Series V. Science and Technology Policy – Vol. 51 ISSN: 1387-6708

Science and Technology Policies for the Anti-Terrorism Era

Edited by

Andrew D. James

PREST – Policy Research in Engineering, Science & Technology
University of Manchester, UK

IOS
Press

Amsterdam • Berlin • Oxford • Tokyo • Washington, DC

Published in cooperation with NATO Public Diplomacy Division

Proceedings of the NATO Advanced Research Workshop on
Science & Technology Policies for the Anti-Terrorism Era
Manchester, United Kingdom
12–14 September 2004

ISBN 1-58603-646-7
Library of Congress Control Number: 2006929619

Publisher
IOS Press
Nieuwe Hemweg 6B
1013 BG Amsterdam
Netherlands
fax: +31 20 687 0019
e-mail: order@iospress.nl

Distributor in the UK and Ireland
Gazelle Books Services Ltd.
White Cross Mills
Hightown
Lancaster LA1 4XS
United Kingdom
fax: +44 1524 63232
e-mail: sales@gazellebooks.co.uk

Distributor in the USA and Canada
IOS Press, Inc.
4502 Rachael Manor Drive
Fairfax, VA 22032
USA
fax: +1 703 323 3668
e-mail: iosbooks@iospress.com

PRINTED IN THE NETHERLANDS

Science and Technology Policies for the Anti-Terrorism Era
Andrew D. James (Ed.)
IOS Press, 2006

Preface and Acknowledgements

This book is the product of a NATO Advanced Research Workshop held in Manchester, UK from 12th to 14th September 2004. The Workshop was hosted by the University of Manchester's PREST science and technology policy research centre. The Workshop was directed by Andrew James of PREST and co-directed by Professor Myklov Ozheven, Kiev Taras Shevchenko National University, Ukraine.

The objectives of the Workshop were to critically consider the science and technology policies necessary for defence against terrorism and other threats to security; to assess the priorities for governments, universities, national laboratories and industrial firms; to identify how governments and the science and technology community can most effectively work together to enhance our security; and to share the experiences of policy makers and policy analysts.

The importance and relevance of the Workshop topic to the policy community was reflected in the seniority of the speakers and participants. These included Dr Parney Albright who at the time of the Workshop held the position of Assistant Secretary for Science & Technology at the U.S. Department of Homeland Security as well as senior figures from the UK Home Office, UK Office of Science & Technology, the European Commission and NATO. Our Key Note Speaker was Dr Ian Gibson MP who at the time was Chairman of the House of Commons Science & Technology Committee. The timeliness of the Workshop was also reflected in the high quality of participants from the academic and think tank communities. In all, 50 people attended the Workshop. Participants were drawn from 11 countries representing NATO, the Partnership for Peace countries and the NATO Mediterranean Dialogue.

Active and open discussion and the sharing of experience were critical to the success of the Workshop and I would like to thank all the participants for their contributions. I would also like to thank several colleagues at PREST without whom the Workshop would have been impossible. In particular, Jessica Chen, Deborah Cox and Susan Nicholson provided excellent organisational support. James Allen's editorial assistance is also gratefully acknowledged.

Andrew D. James
Manchester, UK

June 2006

Contents

Part 1

Introduction

Science and Technology Policies for the Anti-Terrorism Era
Andrew D. James (Ed.)
IOS Press, 2006

Science and Technology Policies for the Anti-Terrorism Era

Andrew D. JAMES
PREST, University of Manchester,
Oxford Road, Manchester M13 9PL, United Kingdom

Abstract. This Chapter introduces some of the key questions that are addressed by the authors within this book. What do we mean by a "new anti-terrorism era" and what, if anything, is new about the security challenges that we face today? What contribution can science and technology make to counter terrorism? What are the public policy responses and how should they be judged? What is the extent of international cooperation in anti-terrorism science and technology development? What are the implications of the new security environment for the governance of science and technology? The answers to these questions are critical to formulating sound science and technology policies to confront one of the key international security challenges of our era.

1. Introduction

There are those who believe that everything changed on September 11th 2001. As U.S. President George W. Bush said of that day: "night fell on a different world" [1]. Bombings in Bali, Madrid, London and elsewhere have reinforced the sense of change in the international security environment. The world, it is argued, has entered a new era and policy attention has increasingly shifted from the threat of conventional inter-state violence to the threat of international terrorism.

In this new anti-terrorism era, science and technology is simultaneously seen as part of the threat but also a key element in the security response. The possibility that terrorists may seek to use a chemical, biological, radiological or nuclear (CBRN) device is being taken seriously by policy makers. At the same time, science and technology is seen as having an important role in support of counter-terrorism strategy

Whilst this new challenge to international security is profound, there are many who rightly caution that overstating the threat is a sure way to costly and ill-judged public policy responses. There are those who look particularly at the United States and question whether the huge scale of the scientific and technological resources being deployed in support of the "homeland security" mission is proportionate to the threat. Certainly, the European response has been more cautious. In part, this reflects differences in assessment of the threat but it also reflects the fact that Europe has many decades experience dealing with terrorism and countries such as the United Kingdom already have a well developed scientific and technological infrastructure directed at conventional terrorism.

Nonetheless, there is no doubt that this new anti-terrorism era poses considerable challenges for science and technology policy on both sides of the Atlantic. The contributions to this book are divided into four broad themes that reflect those challenges. The first challenge is to ensure that science and technology has an appropriate role in counter terrorism strategy. The status of scientific advice in the policy making process and accurate assessment of risk are important issues in an era where scientific and technological threats are deemed important. Equally, how technologies are developed and integrated into counter terrorism strategy will influence our future security. The second challenge is to develop effective public policy responses that engage the science and technology community. The third challenge is to expand effective international cooperation in counter terrorism science and technology efforts. The fourth and final challenge is to ensure that the new security environment does not overly impact the conduct of science and international scientific cooperation.

2. A New Anti-Terrorism Era?

2.1 The New Terrorism

What do we mean by "a new anti-terrorism era" and what, if anything, is new about the security challenges that we face today? After all, terrorism was an ever present feature of the security landscape in Europe and the Middle East throughout the second half of the 20th century. Organised groups with relatively clear political, social or economic objectives sought to use violence as a way of gaining media attention for their cause or to create such disruption and fear in a society that policy decisions were influenced in their favour. This political terrorism was exemplified by radical leftist groups such as the Japanese Red Army, the Red Army Faction in Germany and the Red Brigades in Italy; nationalist separatist movements such as the Provisional IRA and ETA; and Palestinian terrorist groups such as Black September and the Abu Nidal Organisation [2]. Terrorist acts tended to focus on plane hijackings, embassy bombings and attacks on symbolic figures such as soldiers, police, judges and politicians. The assumption of the security community was that, as Brian Jenkins has put it: "Terrorists want many witnesses, not many dead" [3: p.46].

The new terrorism, however, is argued to be different in its aims; targets; and means [4, 5, 6, 7]. Organised groups with clearly defined political motives are argued to have been replaced by a loosely connected network of individuals and groups motivated by miltant Islamism and a common rejection of Western modernism that are seeking to so damage Western society as to "destroy" it in some sense [8]. The London suicide bombings in July 2005 and the Madrid bombings of 2004 graphically and terribly illustrate many of the characteristics of the new terrorism. These were attacks conducted by loosely structured groups whose choice of targets and attack methods were specifically designed to maximise the number of civilian deaths and injuries. There were no coded warnings; no opportunities to clear the area; and, in the case of the London bombings, the bombers were willing to commit suicide in executing the attacks.

2.2 The Spectre of Catastrophic Terrorism

These terrorist attacks have involved the use of conventional explosives. The materials used in the London bombings, for instance, were readily and commercially available and were legal to possess until assembled into the bombs themselves. It is likely that high explosives will remain the weapon of choice for terrorists, nevertheless, the possibility that terrorists may seek new kinds of targets or use new kinds of weapons to inflict death and destruction on a catastrophic scale has moved up the policy agenda.

Concerns have been raised as to the vulnerability to terrorist attack of nuclear power plants, chemical plants and stores of toxic or flammable materials, as well as human and agricultural health systems [9]. There are also concerns that terrorists may seek to exploit vulnerabilities in information technology, transport, power and other critical infrastructures [10, 11].

The possibility that terrorists may seek to acquire and use a chemical, biological, radiological or nuclear (CBRN) device is being taken seriously by policy makers although it ought to be noted that such concerns pre-date the events of September 11th 2001. The potential threat of catastrophic or super terrorism using a CBRN device was the subject of debate during the 1990s and this was prompted by a number of developments including the sarin gas attack on the Tokyo subway by the Aum Shrinkyo cult, fears about the security of nuclear weapons in the former Soviet Union as well as new information on the extent of the Soviet era bioweapons programme [4, 6, 7]. The fear of nuclear or radiological smuggling from the former Soviet Union remains great whilst the anthrax postal attacks in the United States in 2001 gave greater credence to the potential use of biological agents by terrorists. Scientific developments in the life sciences and their "dual use" nature have been the subject of growing concern [12, 13, 14, 15].

The United States has responded to this perceived threat with a massive programme of spending on CBRN countermeasures especially in the bioterrorism field. The UK has established a more modest civilian CBRN research programme designed to be proportionate the UK government's assessment of the threat and, like other Western governments, it is stockpiling anti-viral drugs. Concern about the threat of dual-use technology diffusion has led to new restrictions on universities and foreign students in the United States.

Several chapters in this book address such issues. In his Chapter, Alistair Hay considers the challenges to the emergency services and others if confronted by a chemical or biological incident and the role that the scientific community can play in helping develop equipment, measurement tools and protocols. At the same time, he also argues that the threat of a CBRN terrorist attack should be kept in perspective and that the technical challenges of mounting a CBRN attack are such that the likely scale of such an attack is less than the fear it engenders. In his Chapter, Bill Durodié argues that the very fact that scientists and policy makers are engaged in this kind of public discussion can somehow make that threat of catastrophic terrorism seem more "real" and more "likely" in the minds of the public. This raises anxieties and fears that are out of proportion to the possibility of such an event. In its Chapter, the Royal Society emphasises the individual and collective roles of scientists in ensuring that biomedical science is not misused for terrorist purposes. Picking up on the dual-use nature of the life sciences, Brian Rappert's Chapter considers ongoing attempts to balance security and openness in the conduct of civilian bioscience and biomedical research. Al Teich chronicles how new laws related to foreign students and visitors, access to certain chemical and biological agents in the laboratory as well as the publication of research results and the tightened enforcement of other laws since the September 11th attacks have affected scientific research and higher education in the United States.

2.3 Coming to a Realistic Assessment of Risk

Policy makers face a number of challenges in structuring their response to the potential threat of catastrophic terrorism. Simply put, the risk of a catastrophic CBRN attack is generally thought to be low but the potential consequences could be truly enormous. What should public policy makers do about it? How much public money should we allocate to CBRN countermeasures and preparedness relative to responses to conventional terrorism? Are the resources that are being devoted to the threat of CBRN terrorism, particularly in the United States, proportionate to the threat?

The dilemma faced by policy makers is neatly summed-up in the Chapter by Parney Albright and Holly Dockery of the U.S. Department of Homeland Security (DHS). Albright and Dockery observe:

> "the CBRN threat presents nearly unimaginable consequences. However, the classic risk calculus does not apply. Risk is normally thought of as the product of probability of occurrence multiplied by the consequences. However one interprets probability of occurrence (e.g. some measure of the degree of difficulty for the threat to mount the attack), in some cases that 'measure of belief' about the threat will be judged to be extremely low (e.g. the nuclear threat), while the consequences extremely high. In effect, one is forming a judgment based (qualitatively) on multiplying zero times infinity.... Ultimately, those decisions will instead devolve to how a decision maker 'feels' about the risk from the threat, and on the expected costs and efficacy of the investment decision".

There is another reason, in this context, why conventional risk analysis (probability multiplied by impact) may be meaningless. Probability implies blind and unresponsive chance, yet terrorists are likely to respond to the counter-terrorism measures put in place by governments: this is a two player game, not a game against nature. Terrorism risk is not caused by an exogenous event such as an earthquake or human error but by the deliberate action of individuals seeking to exploit vulnerabilities in counter terrorism security measures [16]. There are many in the policy community who believe that, ultimately, government has to assess priorities on a basis that is at least partly vulnerability driven, not just risk driven. In other words, one relevant test is whether, in the event that a certain kind of vulnerability in the national framework was successfully attacked, would the government be open to the twin accusations: you should have seen that such an attack was possible, and the cost of protecting against it would not have been disproportionate, given the damage and casualties that have resulted.[1]

3. The Contribution of Science and Technology to Counter Terrorism

The role that science and technology may play in counter terrorism is an important question and is the focus of Part 2 of this book. Two types of contribution can be identified. First, the scientific community can provide scientific advice to public policy makers. Second, technology can be used in support of counter-terrorism strategy to pursue and disrupt terrorists, protect against attack and aid in the response and recovery from incidents.

3.1 Scientific Advice to Government

In the CBRN area and elsewhere, the scientific community has an important role to play in establishing a scientific evidence base to inform and support policy and planning decisions. In the UK, the Home Office leads the government's response to counter-terrorism. The Home Office has a CBRN Team and a Terrorism Protection Unit with scientific secondees from the Home Office Scientific Development Branch and the Ministry of Defence's Defence Science and Technology Laboratory (DSTL). The Ministry of Defence advises on technology development and basic science through the DSTL. An ad hoc independent advisory committee called the Scientific Advisory Panel for Emergency Response (SAPER) advises on longer term strategic scientific issues for emergency response, with members drawn from government departments and academia. The Chief Scientific Adviser to the Government also provides scientific advice along with Departmental Chief Scientists from the Home Office, Department of Health and so forth.

Such structures for providing scientific advice to government raise important questions. What knowledge do users need? What are the challenges and opportunities for improving the link between the expertise of the science and technology community and the needs of decision makers? In their Chapter, James Richardson, Whitney Matson and Robert Peters consider the status of scientific advice in the U.S. policy making process and argue that there is a need for better decision processes on national security issues with a science and technology dimension. Too often at the top levels of government there is no "scientist at the table" to take part in critical discussions. Instead, they argue, technical questions are more likely to be judged on the basis of political ideology rather than scientific metrics and "ideology and short-term fixes are too often substituted for depth and objectivity". If Richardson and his colleagues are correct then there are good reasons for anxiety about the quality of decision making on scientific and technological issues within counter-terrorism policy.

What of the contribution of the broader scientific community? In the aftermath of the September 11th attacks the U.S. science and engineering community responded in a variety of ways. One of the most striking was the initiative of the National Academies. Expert working groups brought together many of the leading figures in U.S. science and engineering to consider the scientific and technological threats and opportunities faced by the United States and the role of science and technology in combating terrorism. The outcome was a report *Making our Nation Safer – the Role of Science and Technology in Countering Terrorism*.

Similarly, the UK's national academy of science, the Royal Society, has produced a series of reports on the scientific aspects of international terrorism. The findings of two of those Royal Society reports are discussed in this volume. In his Chapter, Alastair Hay discusses the work of a Royal Society working group on detecting and decontaminating chemical and biological agents and the main findings of its report *Making the UK Safer*. A separate Chapter summarises the Royal Society's views on the roles and responsibilities of the scientific research community with regard to the proliferation and use of CBRN weapons. This is set out in its statement *The Individual and Collective Roles Scientists can Play in Strengthening International Treaties*. In his Chapter, Bill Durodié takes issue with the nature of these studies. His criticism is directed against the Royal Society's report *Making the UK Safer* but could equally be directed at the U.S. National Academies study. Durodié's concern is that such reports take at face value the underlying political and public policy assumptions of our age. Namely, that the threat of the use of chemical and biological agents against civilian targets is indeed greater following September 11th and that science and technology are central to reducing the threat. In Durodié's view, the failure to question the underlying premises of the debate is dangerous and he argues that: "It is not just the job of social scientists, but scientists too, to question whether this purported 'greater prominence' is real".

3.2 Technological Responses to Terrorism

A host of technologies are being developed and deployed to aid the pursuit and disruption of terrorist organisations; the protection of critical national infrastructure and key sites; and to aid response and recovery from terrorist attack. A number of studies have sought to identify the potential contribution of existing or future technologies [9, 17, 18, 19, 20].

Table 1: Some examples of the application of technology to counter-terrorism strategy

	Elements of counter-terrorism strategy			
	Pursuit	**Protection**	**Preparedness**	
			Response	**Recovery**
Aim	Understand better the capabilities & intentions of terrorist organisations. Disrupt their ability to operate at home & overseas.	Protection of areas at particular risk including the critical national infrastructure & potential terrorist targets.	Develop contingency arrangements for response to a terrorist attack.	Develop contingency arrangements for recovery from a terrorist attack.
Missions	Intelligence gathering. Surveillance. Border security.	Detection of persons, explosives & CBRN materials. Protection of critical infrastructures. Transport & port security.	Emergency management. Effective first response by police, ambulance & fire services. Detection, diagnosis & treatment of CBRN agents. Neutralisation & containment of the effects of a terrorist attack.	CBRN decontamination of people, buildings & infrastructure.
Examples of the use of science & technology	Data fusion, data storage & data mining. Identification of suspicious patterns of behaviour from databases or sensors. National identification card systems. Biometric ID.	Sensors. Fire walling & virus protection. Encryption & trusted computing. Non-intrusive cargo screening for ports of entry. RFID technologies to track & authenticate cargo.	CBRN sensor detection systems. CBRN point detectors for first responders. Interoperable communications systems for first responders. Personal protection equipment for first responders. Research on treatments & preventatives for known & potential pathogens. Modelling of effects of CBRN attack.	Improved methods & standards for decontamination.

Table 1 provides examples from these studies and shows how these technologies can support the different missions faced by a counter-terrorism strategy.

What ought to be emphasised here is that many of the technical countermeasures identified in these studies are based on existing technologies that can find immediate applications. They are often already in use in the defence sector, the security sector or elsewhere. The challenge for governments is to provide funding for development and procurement to pull through such technologies into the hands of users. Examples include security systems for shipping containers, the production and distribution of known treatments and preventatives for pathogens or interoperable communications systems for first responders [9]. Others require more extensive scientific research and will only be available for use in the longer term. Examples include advances in data fusion and data mining for intelligence analysis and the development of improved CBRN sensor and surveillance systems for use by emergency officials and first responders [9].

In their Chapter, Parney Albright and Holly Dockery emphasise that there may be greater and quicker pay-backs by funding the adaptation of technologies that have been developed for other purposes and often for civilian applications. They note that this presents challenges to the U.S. Department of Homeland Security as it seeks to develop a balanced R&D portfolio. In particular, it raises questions about the appropriate balance between development funding aimed at adapting existing technologies for the anti-terrorism mission and funding research that may lead to longer-term high impact pay-backs.

Three chapters in this volume consider potential technological responses to terrorism. In their Chapter, Isaac Ben-Israel, Oren Setter and Asher Tishler argue that the emphasis of anti-terrorism R&D should be on what they call long-term intelligence (LTI) systems. LTI systems place an emphasis on the development of more sophisticated automatic, computer-based management information systems to analyse data over long periods of time. There is a need for data collection from civilian as well as defence sources and enhanced data fusion, data storage and data mining methods and methodologies to identify "alarming behaviour". All these should be developed whilst protecting civil liberties. In his Chapter, Alastair Hay who was a member of the Royal Society working group on chemical and biological detection systems summarises the key themes of the resulting Royal Society report. *Making the Nation Safer* identified need for further work on chemical and biological detection systems with the objectives of: exploiting new and existing science, engineering and technology for robust detection of chemical and biological agents; developing point detectors for first responders; establishing scientific data on the reliability and sensitivity of different detection systems under different background conditions; and, the analysis of medical intelligence to enhance effective response. Finally, in their Chapter, Konstantin Volchek and Merv Fingas discuss the technologies necessary for decontamination after a CBRN terrorist event and the challenges of developing adequate, reliable and economic decontamination methods. They emphasise the need for improved detection methods and personnel protection equipment.

4. Challenges for Counter-Terrorism Science and Technology Policy

Counter-terrorism science and technology policy is being developed in a multi-agency environment with a diverse range of technology users and diverse sources of technological knowledge and capability. This poses particular challenges for public policy makers and some of those challenges are addressed in Part 3 of this book.

4.1 Capturing User Requirements

Users of counter terrorism technologies include a diverse range of public and private sector organisations with an equally diverse range of missions and capabilities. In Europe, users range from border guards on the Polish-Ukrainian border concerned with people trafficking to the Metropolitan Police in central London dealing with the reality of suicide bomb attacks. They stretch from those charged with the security of major public events like the Athens Olympics or the football World Cup in Germany to the fire and rescue services and those responsible for security at ports and airports. Within government, counter terrorism tends to cross-cut traditional departmental boundaries and counter terrorism and resilience is the responsibility of not only national but also local government. At the same time, most of the critical infrastructure is owned by the private sector.

Thus, how to effectively capture user requirements is an important issue for those responsible for counter-terrorism science and technology development. Effective priority-setting, user engagement and good management are all important. In their Chapter, Holly Dockery and Parney Albright discuss how priorities are being set within the DHS science and technology programme. The emphasis is on a strategic approach based on a constant examination of U.S. vulnerabilities and evaluation of the threat and its weaknesses. Dockery and Albright stress the importance of user engagement at all stages of the research, development, testing and evaluation cycle. To be effective, they argue, the process has to move away from "technology push" to ensure that the focus is on the highest priority issues and that user capability needs are adequately captured. In his Chapter, Cam Boulet discusses how the Canadian government has also engaged in a capability analysis process. The exercise, begun prior to September 11th 2001, sought to identify priority capability requirements through the use of a Consolidated Risk Assessment methodology. The CRA methodology used scenarios of CBRN events to evaluate public safety vulnerabilities and derive science and technology programme priorities.

4.2 Transferring Defence Capabilities for Civilian Use

A second significant public policy challenge is how to make best use of established defence-related scientific and technological capabilities. In the UK, the Home Office may have formal responsibility for civilian counter terrorism science and technology but it draws on the scientific and technological capabilities of the MOD, Department of Health and many other departments. Particularly in the CBRN field, most of the government R&D capabilities relevant to counter-terrorism have historically resided in the defence research establishments. Indeed, the House of Commons Science and Technology Committee has argued that UK home defence is too dependent upon these military-derived technologies and expertise based in government defence laboratories [21].

In his Chapter, Alastair Hay notes that the Royal Society has argued that whilst the UK has a great deal of knowledge and expertise in the CBRN area, no single government department appears to have full responsibility for determining how this expertise can be best utilised. In their Chapter, Parney Albright and Holly Dockery describe the rationale for the Science and Technology Directorate within the new U.S. Department of Homeland Security. Significantly, the Science and Technology Directorate may have responsibility for science and technology policy in the counter-terrorism field but critical scientific and technological capabilities reside in the Department of Defense, the National Institutes of Health and other organisations such as the Department of Agriculture.

The challenge of how to make best use of defence capabilities extends beyond scientific advice. There is also the question of the extent to which established defence-related technologies can be used for civilian counter-terrorism missions. Technologies developed for the military are not necessarily immediately transferable to civilian counter-terrorism needs and, in this book, this point is raised in the Chapter by Heiko Borchert and is also discussed at some length by Parney Albright and Holly Dockery. The technology needs of the military mission, Albright and Dockery argue, are very different to those required by civilian first responders and other users of counter terrorism technologies. Equipment developed for the military assumes a level of training, technical support and so forth that may be impractical for civilian organisations. Military equipment is designed to be used intensely but periodically whilst civilian requirements are for constant use with a minimum of false alarms or false positives. At the same time, use in the civilian environment raises important issues related to legal liability and insurance risk.

4.3 Engaging the Wider Science and Technology Community

A further challenge is how to engage the wider science and technology community. There is a recognition that scientific and technological knowledge that is important to counter terrorism may lie outside the traditional military-industrial-technological complex. Networking, Information and Communication Technologies, software and knowledge management systems, biotechnology and robotics are just a few examples of the commercial (non-defence origin) technologies that have been identified as potentially important to counter terrorism [9]. Reaching-out to these non-traditional sources of scientific and technological expertise presents a considerable challenge. In their Chapter, Parney Albright and Holly Dockery emphasise the importance that the Department of Homeland Security places on mobilising the U.S. scientific community whether in the private sector, government labs or universities. A Homeland Security Centers of Excellence Program has been established to engage the academic community by establishing university-based centres of multi-disciplinary research in areas such as agricultural, chemical, biological, nuclear and radiological, explosive and cyber terrorism and the behavioural aspects of terrorism. The DHS University Program is seeking to "foster a homeland security culture within the academic community through research and educational programs" and to "strengthen U.S. scientific leadership in homeland security research".[2]

However, major challenges have been encountered. Summing up the situation two years after 9/11, an editorial in *Nature* declared that: "the promised reorientation of U.S. national research priorities proceeds without much direction or conviction.... Two years on, that expertise has been tapped only sparingly" [22: p.107.]. Lewis Branscomb has noted how retaining political momentum and public interest in science and technology policies for anti-terrorism may be a challenge – especially if there are no further attacks on the U.S. homeland [23]. There is no doubt that academic scientists in the United States have enthusiastically pursued the new research funding opportunities. In the wider context, however, the relationship between the science community and the Bush Administration has been strained by the conduct of the war in Iraq as well as the consequences of Administration policies for the conduct of science in the United States.

Some of the challenges of engaging the science and technology community are addressed in the Chapter by Heiko Borchert. Borchert argues that the complex interplay between the normative, regulatory, economic and scientific incentives and disincentives faced by actors will determine their willingness to engage with the new anti-terrorism agenda. The Canadian experience may provide a pointer here. In his Chapter, Cam Boulet describes how the Canadian CBRN Research & Technology Initiative has placed an emphasis on collaboration across government agencies and openness to all sectors of the national innovation system. What Boulet describes as a "strong community building approach" involves a concerted effort to reach out to private, academic and end user communities. In his Chapter, Alastair Hay emphasises the importance of coordination and organisation if the UK is to successfully utilise the full range of its capabilities in government, industry and universities. He observes that developments may well occur in research disciplines that have not traditionally been focused on military or security related matters and that means are needed to alert these academics to the relevance of their work. Hay sets out the Royal Society's recommendations for policy initiatives in the UK, including a new centre to coordinate and direct UK efforts in this field and to determine, commission and direct civilian planning, preparedness and R&D efforts in the field of decontamination and detection. The House of Commons Science and Technology Committee pursued a similar theme, arguing that there was a need for a Centre for Home Defence under the Home Office to develop technologies for civil use and provide the hub of a national network of researchers to tap into research capabilities in universities and elsewhere [21].[3] There have been some efforts to engage new sources of knowledge in the UK. The Home Office has launched a CBRN Science and Technology Programme that aims to make sure that there is a firm scientific basis for planning the protection of the UK from a CBRN terrorism incident. This has sought to engage private industry and universities.

4.4 Creating a Public-Private Partnership

A further critical challenge is how to build an effective partnership between the public and private sectors. Most critical infrastructure and many potential terrorist targets are in private sector hands and investments in their protection are highly dependent upon the decisions of private sector organisations. The same is true of resilience planning against the consequences of terrorist actions. How to motivate private investment in hardening critical infrastructure has become a major public policy issue and there is much discussion of so-called "civilian benefit strategies" – ways of hardening critical infrastructures whilst generating wider commercial benefits to the private sector [23]. In their Chapter, Holly Dockery and Parney Albright observe that many of the new homeland security systems have the potential to provide private sector benefits: container tracking technologies, for instance, may enhance the efficiency with which we can transport goods. At the same time, they observe, the private sector is unlikely to adopt new security technologies if they impede commerce and impose additional costs on business.

Equally, there are important challenges in engaging private sector companies in the market for counter terrorism technologies. The nature of government procurement contracts and the relatively small size of the markets for security-related applications of their technology often make the emerging counter terrorism market unattractive to non-defence contractors. Unlike the defence equipment sector, national governments have little leverage over some of the key sectors deemed important for the anti-terrorism mission. A good example is the pharmaceutical sector where the limited scale of the market for vaccines and drugs to counter bioterrorist agents makes it a commercially unattractive area for large scale private sector R&D investment [9, 21, 24].

These and other economic considerations are discussed in the Chapter by Heiko Borchert. He considers how governments can provide incentives for private sector investment and he reviews the U.S. government's Project Bioshield. This is not an R&D programme but is designed to encourage private sector R&D investments in biodefence vaccines, therapeutics and other countermeasures. Project Bioshield aims to provide a guaranteed government market for future products through procurement and stockpiling of drugs.

5. Enhancing International Cooperation

Another issue of concern to many policy makers is how to enhance international cooperation in counter-terrorism science and technology. This is the subject of Part 4 of this book.

There is some transatlantic cooperative activity on a bi-lateral government-to-government basis and through NATO and there have been some discussions between the U.S. Department of Homeland Security and the European Commission. An example of bi-lateral cooperation is the Memorandum of Agreement between the UK and U.S. governments signed in 2004. *Co-operation in Science and Technology for Critical Infrastructure Protection and Other Homeland/Civil Security Matters* establishes a framework for future science and technology co-operation to promote UK-U.S. counter-terrorism research and to seek the best expertise available to carry out a joint science and technology programme. The Memorandum of Agreement began a process of agreeing areas of work, joint requirements for each area, and each country's responsibilities for delivering their parts of the programme [25]. There have been prior cooperative agreements including a Memorandum of Understanding on Counter Terrorism Research and Development signed between the Ministry of Defence and the Department of Defense in 1995. UK-U.S. cooperation on nuclear, biological and chemical countermeasures has a long history.

Neither Europe nor the United States can afford wasteful duplication in this field and improved coordination of this kind makes a great deal of sense. Yet transatlantic cooperation on military science and technology and weapons programmes has been fraught with difficulties. Is transatlantic cooperation on anti-terrorism technologies likely to be any more successful? The Chapter by Holly Dockery and Parney Albright suggests that it might. They set out the case for international cooperation and make it clear that the Department of Homeland Security sees many benefits. Even with its vast scientific and technological resources, they say, the United States is still stretched and overcoming capacity constraints in areas such as WMD research is important. Homeland security science and technology is different to military science and technology and more amenable to international cooperation and the Canada-U.S. Public Security Technical Program is put forward as a template for international science and technology cooperation. Dick Bitzinger is less hopeful for transatlantic cooperation. In his Chapter he sets out many of the challenges that have been faced in transatlantic cooperation on military R&D and weapons programmes including the nature of U.S. export controls, controls over security of supply and information as well as a U.S. proclivity towards protectionism. He argues that efforts by the U.S. Congress to impose "Buy American" provisions on homeland security technologies may also present a stumbling block to transatlantic cooperation.

In his Chapter, Heiko Borchert discusses developments at the European level and in particular the European Commission's initiative in security research. The European Commission has traditionally been excluded from direct involvement in defence R&D but is seeking to play a role in counter terrorism R&D through the European Security Research Programme. The political and institutional significance of the European Security Research Programme should not be underestimated. This is the first major initiative of the European Commission in the security field and, if it is to work, it will require the development of new relationships between national and local agencies, the science and technology community in Europe and the European Commission. If the Commission's initiative achieves its aim of enhancing coordination and reducing duplication then it could have a significant impact on

Europe's technological capabilities in the anti-terrorism field. However, how the Commission is going to pull through technologies funded under the European Security Research Programme into the hands of national and locally-organised first responders remains a key issue [26].

6. The Consequences of the New Security Environment for the Conduct of Science

The consequences of the new security environment for the governance of science and technology have been the subject of considerable debate within the science community and this is the topic of the Chapters in Part 5. The controversies that have emerged regarding scientific freedom, the role of universities and so forth hark back to Cold War concerns within the scientific community about the implications of military funding of scientific research: its potential to distort scientific priorities and the course of scientific and technological developments; constrain the scientific freedoms of those engaged in such funded research; and its impact on the nature of research universities [27]. However, the pervasive nature of dual use technologies gives these controversies a new edge not least in the life sciences.

6.1 Controls on the Conduct of Scientific Research

Since the 9/11 attacks, the U.S. government in particular has introduced new controls that have had implications for the conduct of scientific research in the U.S. In his Chapter, Al Teich discusses how new laws such as the USA PATRIOT Act as well as tougher enforcement of existing regulations have had an impact on scientific research and higher education in the United States. Foreign student numbers have fallen; new regulations have been put in place governing controls over laboratory use of chemical and biological agents; and some government agencies have sought to control the dissemination of so-called "sensitive but unclassified" information. There have been warnings from the U.S. scientific community that this is hampering international scientific cooperation. There have also been warnings that the new security regulations may stifle creativity, drive the best scientists away from research in sensitive fields and ultimately weaken the anti-terrorism science and technology effort [24, 28]. There are concerns about what such developments may mean for the research universities. Fears have been expressed that the new security environment presents risks to research universities by placing limitations on researchers' access to data, challenging their commitment to openness and the free exchange of information and imposing controls on foreign students. Eugene Skolnikoff warns that:

> "I think we may have serious trouble ahead, and time is short. Obviously, the universities are moving in one direction, toward greater internationalization while jealously guarding the essential openness of the campus. At the same time our national concerns about proliferation, movement of information, and access of foreign students are intensifying in the opposite direction" [29: p.71].

Most of the attention has focused on the United States and there has been less public debate about such matters in Europe. In the UK, the Anti-Terrorism, Crime and Security Act 2001 (ATCSA) introduced new controls aimed at tightening the security of pathogens and toxins held in laboratories and made it an offence to aid or abet the overseas use or development of chemical, biological, radiological or nuclear weapons. The Export Control

Act 2002 provides the Government with powers to impose controls on the transfer of military and dual-use technologies by intangible means and during the passage of the Bill there were concerns from the scientific community that it could potentially stifle the flow of scientific knowledge and hinder international collaboration in some fields. A Voluntary Vetting Scheme has been in place for some time under which universities are invited, on a voluntary basis, to refer to the Government for advice applications from students from certain countries in certain disciplines. The Voluntary Vetting Scheme was established in 1994 and is aimed at preventing the transfer of technology related to weapons of mass destruction. The operation of the Voluntary Vetting Scheme was the subject of criticism by a report of the House of Commons Foreign Affairs Committee and at the time of writing it was being reviewed [30].

By and large, such matters have been sensitively handled by the UK government, the police and security services and there have been little of the kinds of debates and problems that have arisen within the U.S. scientific community. One study suggests that the implementation of biosecurity controls in the UK has had a limited negative impact on the UK scientific community and has been less disruptive in the UK than in the U.S. or Germany [31]. However, there have been warnings issued to universities that they need to be aware of the new security environment. The Foreign Affairs Select Committee called for tighter restrictions on university bioscience research. Indeed, the then Chairman of the House of Commons Science and Technology Committee has warned that: “The scientific community needs to take stronger action to regulate itself – otherwise it may have ill-judged restrictions placed on it by politicians”[4].

6.2 The Moral and Ethical Responsibilities of Scientists and Engineers

The Royal Society and the Wellcome Trust have sought to stimulate awareness and debate on such issues within the life science research community in the UK and have discussed amongst other matters how research institutions and funding bodies can better review research projects to minimise the risk of misuse; the communication and publication of potentially sensitive research results; and education and awareness raising of the ethical and legal responsibilities of the scientific community [28, 32]. Malcolm Dando observes that:

> “The problem for many life scientists is that they are not conversant with the issues surrounding biological weapons, biowarfare, bioterrorism and associated topics, despite almost a century of biology and medicine being used in offensive bioweapons programmes” [33: p.149]

The need to raise awareness amongst the research community of the ethical and legal aspects of their work has been the topic of a Royal Society paper. That paper on the *Individual and Collective Roles Scientists Can Play in Strengthening International Treaties* is reproduced as a Chapter in this book. The Royal Society observes that the scientific community must exercise judgement in the publication of its work and that there is a need to raise awareness within the research community of its responsibilities. The Royal Society raises the subject of Codes of Conduct, particularly for life scientists, and argues that the aim should be to define enforceable Codes governing ethical and moral aspects related to good conduct in life science research, including the potential misuse of results for illegal weapons development.

The issue of Codes of Conduct is one of the topics of the Chapter by Brian Rappert who assesses on-going attempts to balance security and openness in the conduct of civilian bioscience and biomedical research. Rappert argues that current debates and policy prescriptions are unclear as to the ultimate goals of controls, the desirability of the diffusion of dual use knowledge and the matter of who should decide what constitutes "dangerous" research. Uncertainty and disagreement exists regarding the way forward and Codes of Conduct for life scientists have both potential and pitfalls.

6.3 A Distortion of Scientific Priorities?

During the Cold War, military funding of scientific research prompted concerns amongst those who feared its potential to distort scientific priorities and the course of scientific development. With the rapid growth of counter-terrorism related research funding, such concerns are being voiced once again and not least by the U.S. microbiology community. Concerns have been expressed that the growth in biodefence funding may well alter the direction of training and research in the life sciences and that the increased emphasis on biodefence may change the character of scientific meetings and publications. Indeed, the American Society of Microbiology has already decided to host a separate annual biodefence research conference in part because of concerns that other general meetings would be overwhelmed and biased by the number of biodefence papers. As a leading figure in the U.S. microbiology community has observed:

> "Abundant new funds are available for biodefence research, and many researchers are racing to enter the field... The proposed US biodefence research agenda is likely to change the face of microbiology for many decades" [24: p.70].

This has provoked a lively debate on funding priorities. In March 2005, in an open letter to the National Institutes of Health, some of the leading researchers in the U.S. biological science community declared that:

> "The diversion of funds from projects of high public-health importance to projects of high biodefense but low public-health importance represents a misdirection of NIH priorities and a crisis for NIH-supported microbiological research" [34: p.1409].

These are important issues. There have long been concerns in the U.S. science community that government R&D funding is skewed towards "missiles and medicine" at the expense of other missions and scientific disciplines. Indeed, European scientists and science policy makers would do well to make themselves aware of such controversies as they increase funding for security-related research at the European level.

7. Science and Technology in the Anti-Terrorism Era

There is one further point that ought to be emphasised. This book focuses on science and technology policies for the anti-terrorism era: the role of science and technology in tackling the challenges of the new security environment; national and transnational policy responses; and the implications of the new security environment for the governance of science and technology. These are important issues that deserve extensive discussion. How technologies are developed and integrated into counter-terrorism policy is one important element of our future security. However, there is a danger of seeing science and technology as a panacea and we must remember that scientific and technological responses do not represent "solutions" but merely a response to a particular set of problems. David Lyons observes in his discussion of surveillance technologies:

> "the larger perspective is that 'technology' is still seen as a savior, as the first resort of 'advanced' societies. This is nothing new, but the quest for technologies, geared to guaranteed security, has been gathering pace especially since the end of the second world war. Technological solutions are invoked before other more labor-intensive and human-orientated surveillance methods (which, ironically, are in fact more likely to succeed) let alone efforts aimed at mutual understanding and the reduction of Western threats to Islamic countries" [35: p.667].

We should be highly sceptical of the efficacy of technological solutions to what are ultimately social and political problems. How these problems are defined and understood is critical. Any response to international terrorism needs a broad and contextualised understanding of the root causes of terrorism, an awareness of the consequences of policy actions directed at counter-terrorism and so forth. Science and technology can only be one element in the response to international terrorism. Understanding the underlying causes of terrorism has to be at the heart of our response and an improved understanding of such matters requires the engagement of not only the physical and biological sciences but of the social science community as well.

8. The Structure of the Book

This book is structured as follows. Part 2 considers the role of science and technology in the new security environment. The next two Chapters take very different perspectives on the role of the science and technology community in policy making. Jim Richardson, Whitney Matson and Robert Peters consider the question of how science can be integrated into the policy process. Bill Durodié calls on the science and technology community to take a more critical stance towards the terrorism debate. The following three Chapters consider potential technological responses. Isaac Ben-Israel, Oren Setter and Asher Tishler discuss the use of intelligence technologies to disrupt and apprehend terrorists. Their perspective is strongly informed by the Israeli experience of suicide bombings. Alistair Hay discusses the Royal Society's report on the use of technologies for detection and decontamination of chemical and biological agents and Konstantin Volchek and Merv Fingas focus on the issue of the decontamination after a CBRN event.

Part 3 turns to consider public policy responses. Parney Albright and Holly Dockery discuss R&D within the new U.S. Department of Homeland Security. Cam Boulet describes the Canadian CBRN Research and Technology Initiative. Heiko Borchert discusses the challenges of engaging the science and technology community in such public policy initiatives.

Part 4 considers the potential for greater international cooperation in such matters not least between the United States and Europe. Holly Dockery and Parney Albright discuss how international cooperation fits into the plans of the U.S. Department of Homeland Security. Dick Bitzinger takes a more sceptical line arguing that many of the issues that have dogged transatlantic military technology cooperation could also challenge counter terrorism science and technology efforts.

The final section of the book (Part 5) considers how the new security environment is impacting the governance of science and technology. Al Teich describes the consequences of new security regulations for the conduct of U.S. science and international scientific cooperation. The next Chapter reproduces a paper by the Royal Society that discusses the role of the science community in strengthening international treaties. One issue that the Chapter raises is that of Codes of Conduct for life scientists. This issue is the focus of the Chapter by Brian Rappert, who considers whether dual use means that it is infeasible for the security community to aspire to control the life sciences and critically assesses the notion of Codes of Conduct.

Acknowledgements

I would like to thank Helen Almey, Jonathan Aylen, Shane Brighton and Graham Jordan for their exceptionally helpful comments on earlier drafts of this Chapter. All errors and omissions are, of course, my own.

Notes

1 I am grateful to Graham Jordan for his observations on this point.

2 For further information on the U.S. Department of Homeland Security's University Program see: http://www.dhs.gov/dhspublic/interapp/editorial/editorial_0555.xml

3 In April 2006, the Ministry of Defence announced the establishment of a Counter Terrorism Science & Technology Centre to provide a hub for government laboratories, industry and academia working in the field. The MOD Counter Terrorism Science & Technology Centre will work on science and technology for military and civilian applications.

4 This point was made by the Rt Hon Ian Gibson MP. The scientific response to terrorism. Key Note address to the NATO Advanced Research Workshop on Science and Technology Policies for the Anti-Terrorism Era. PREST, University of Manchester, UK; 2004 September 13.

References

[1] Address to a Joint Session of Congress and the American People. President George W. Bush. 2001 September 20. Available at: http://www.whitehouse.gov/news/releases/2001/09/20010920-8.html

[2] Hoffman B. Terrorism trends and prospects. In: Lesser IO, Hoffman B, Arquilla J, Ronfeldt DF, Zanini M, Jenkins BM, editors. Countering the New Terrorism. Santa Monica: RAND Corporation; 1999.

[3] Jenkins BM. The limits of terror. Harvard International Review 1995; 17 (3): 44-47.

[4] Kupperman R. A dangerous future. Harvard International Review 1995; 17 (3): 46-49.

[5] Laquer W. Postmodern terrorism. Foreign Affairs 1996; 75 (5): 24-36.

[6] Carter A, Deutch J, Zelikow P. Catastrophic terrorism: tackling the new danger. Foreign Affairs 1998; 77 (6): 81.

[7] Lesser IO, Hoffman B, Arquilla J, Ronfeldt DF, Zanini M, Jenkins BM, editors. Countering the New Terrorism. Santa Monica: RAND Corporation; 1999.

[8] Burke J. Al-Qaeda: Casting a Shadow of Terror. London: IB Taurias; 2003.

[9] National Research Council (U.S.) [NRC]. Making the Nation Safer: The Role of Science and Technology in Countering Terrorism. Committee on Science and Technology for Countering Terrorism, NRC. Washington DC: The National Academies Press; 2002.

[10] Homer-Dixon T. The rise of complex terrorism. Foreign Policy 2002; 128: 52-62.

[11] Branscomb LM. The changing relationship between science and government post-September 11. In: Teich AH, Nelson SD, Lita SJ, editors. Science and Technology in a Vulnerable World: Supplement to AAAS Science and Technology Policy Yearbook 2003. Washington DC: American Association for the Advancement of Science, 2003.

[12] Falkenrath RA. Confronting nuclear, biological and chemical terrorism. Survival 1998; 40 (3): 43-65.

[13] National Research Council (U.S.) [NRC]. Biotechnology Research in an Age of Terrorism. Committee on Research Standards and Practice to Prevent the Destructive Application of Biotechnology. Washington DC: The National Academies Press; 2004.

[14] House of Commons Foreign Affairs Committee (UK). Weapons of Mass Destruction. Eighth Report, HC 407 Session 1999-2000; London: The Stationery Office; 2000 July.

[15] Foreign and Commonwealth Office (UK). Strengthening the Biological and Toxin Weapons Convention: Countering the Threat from Biological Weapons. Cm 5484; London: The Stationery Office; 2002 April.

[16] Organisation for Economic Cooperation and Development [OECD]. Emerging Risks in the 21st Century: An Agenda for Action. Paris: OECD; 2003.

[17] Stevens B, editor. The Security Economy. Paris: Organisation of Economic Cooperation and Development; 2004.

[18] The Royal Society (UK). Making the UK Safer: Detecting and Decontaminating Chemical and Biological Agents. Policy Document 06/04. London: The Royal Society, 2004.

[19] Research for a Secure Future. Report of the Group of Personalities in the Field of Security Research. Luxembourg: European Commission; 2004.

[20] Workshop on Research and Technological Challenges in the Field of Border Control in the EU-25 – Proceedings. European Commission Directorate-General for Research, 18-20 October 2004, Ljubljana, Slovenia. Available at: http://europa.eu.int/comm/enterprise/security/doc/proceedings_en.pdf

[21] House of Commons Science and Technology Select Committee (UK). The Scientific Response to Terrorism. Eighth Report, HC 415-I Session 2002-2003; London: The Stationery Office; 2003 October.

[22] Science and the war on terror. Nature 2003; 425 (6954), 11 September: 107.

[23] Branscomb L. Protecting civil society from terrorism: the search for a sustainable strategy. Technology in Society 2004; 26: 271-285.

[24] Atlas RM. Bioterrorism and biodefence research: changing the focus of microbiology. Nature Reviews Microbiology 2003; 1: 70-74.

[25] Co-operation in Science and Technology for Critical Infrastructure Protection and Other Homeland/Civil Security Matters. Agreement between the Government of the United States of America and the Government of the United Kingdom of Great Britain and Northern Ireland. 2004 December. Available at: http://security.homeoffice.gov.uk/news-and-publications1/publication-search/general/st-agreement

[26] James AD. U.S. defence and homeland security spending: an analysis of the impact. Presentation to the Homeland Security Forum Conference, The United States, Europe and Beyond: New Challenges for Institutions and the Private Sector, Geneva Centre for Security Policy, 7-8 October 2004, Geneva.

[27] Smit W. Science, technology, and the military: relations in transition. In: Jasanoff S, Markle GE, Petersen JC, Pinch T, editors. Handbook of Science and Technology Studies. London: Sage; 1995.

[28] Alberts B, May RM. Scientists support for biological weapons control. Science 2002; 298, 8 November: 1135.

[29] Skolnikoff EB. Research universities and national security: can traditional values survive? In: Teich AH, Nelson SD, Lita SJ, editors. Science and Technology in a Vulnerable World: Supplement to AAAS Science and Technology Policy Yearbook 2003. Washington DC: American Association for the Advancement of Science, 2003.

[30] House of Commons Foreign Affairs Select Committee (UK). The Biological Weapons Green Paper. First Report, HC 150 Session 2002-03; London: The Stationery Office; 2002.

[31] McLeish C, Nightingale P. The impact of dual use controls on UK science: results from a pilot study. SPRU Electronic Working Paper Series, Paper no. 132. University of Sussex: SPRU; 2005 April.

[32] The Royal Society (UK). Do No Harm: Reducing the Potential for the Misuse of Life Science Research. Report of a Royal Society-Wellcome Trust meeting held at the Royal Society on 7 October 2004. London: The Royal Society; 2004.

[33] Dando M. The science of life and death. Nature 2004; 432; 11 November: 149.

[34] Altman S, et al. An open letter to Elias Zerhouni, Science 2005; 307; 4 March: 1409.

[35] Lyon D. Technology vs 'Terrorism': circuits of city surveillance since September 11th. International Journal of Urban and Regional Research 2003; 27 (3): 666-678.

Part 2

The Role of Science and Technology

Science and Technology Policies for the Anti-Terrorism Era
Andrew D. James (Ed.)
IOS Press, 2006

Promoting Science and Technology to Serve National Security

James J. RICHARDSON, Whitney MATSON and Robert PETERS
Potomac Institute for Policy Studies,
901 N. Stuart Street, Suite 200, Arlington, VA 22203, USA

Abstract. This Chapter examines the need for better decision processes on national security issues with a science and technology dimension and suggests ways to improve those processes. The Chapter summarises the key points from a study by the Potomac Institute for Policy Studies, titled *Shaping Science and Technology to Serve National Security* and published in greater detail elsewhere [1]. The study was conducted under the auspices of Senators Jeff Bingaman and Joseph Lieberman, Representative Curt Weldon, Deputy Undersecretary for Defense (Science and Technology), Air Force Office of Scientific Research, Department of the Army, Defense Advanced Research Projects Agency, National Intelligence Council, National Science Foundation, Office of Naval Research, and the U.S. Coast Guard.

1. Introduction

Considering the relationship between science and public policy, an article in the journal *Nature* declared:

> " ... the power of science to alter nature has reached such a state that society needs to have a much more fundamental place in considering its support… [and that] … fundamental science is unpredictable, unavoidably sets its own agendas, and has an inherent timescale, both in its community structure and its execution, that is ill matched to the short-term perceptions of public opinion" [2].

Strong science and technology (S&T) bureaucracies, collaborative research both in and out of government, clear channels of information, and adequate S&T budgets are all necessary for an effective S&T programme to counter terrorism. However, these features are not sufficient – they also require good decision and policy-making processes. Producing policy and decisions on S&T matters are particularly difficult and this is at the heart of the article in *Nature*. Since the complexity and effects of scientific breakthroughs and applications will continue to rise over the next twenty years, this difficulty will surely grow unless decision processes are improved. This is important because S&T's immense impacts will fall on both sides of the ledger, some crucially beneficial, others tragically destructive. It is incumbent upon government to pursue decisions and policies that maximise the former and minimise the latter. However, if the fundamental nature of science is so unpredictable and inconsistent with society's range of vision (and attention span), how can government best husband her virtues and control her potential for destruction over the next decades to solve, rather than to create, national security threats? Another point to be made is that more of society's research is being conducted outside of government, so its effects on national security may be less obvious to elected and civil service officials.

These points seem particularly valid because the array of threats to both national and international security has become more diverse in the past decade or so and is likely to continue to become more so. Terrorism is the "threat du jour" and no one can minimise its seriousness, but we cannot turn away from a thousand others that can quickly move to centre stage, from crime and hunger to ecological disaster. If wisely implemented, technology can give us the tools to address many of these concerns in synergistic ways, but that synergism must be articulated in an appropriately broad definition of national security. In this sense, perhaps the Clinton Administration was on the right track in considering the adoption of a broader definition of national security after the Cold War ended.

For example, it will become increasingly challenging to direct our technology development efforts toward providing solutions to scenarios of domestic and international terrorism, while preventing those same technologies from being directed against us, either inadvertently or with malice. For instance, while new sensors are emerging to help in the search for current classes of weapons of mass destruction (WMD), other ongoing research may well deliver mechanisms that are more efficiently destructive than our current crop of WMD, and that resist these sensor solutions. We must remember that these days the enabling research is being conducted worldwide and is not confined to military laboratories.

2. The Science and Technology Challenge for Policy Makers

There are several reasons why S&T issues are being addressed poorly at the top of government in the U.S. To begin with, the technical acumen of politicians in particular, and the populace in general, has not kept pace with S&T progress. Of course, this is mostly due to the enormous growth in scientific principles and applications that have policy implications but it is also a result of the enormous distractions of our age. Many of these have themselves been spawned by advances in S&T not least the remarkable new dimensions added to entertainment through information technology research.

This Chapter concentrates on decision processes that exist at the highest levels of government. In the United States this means Congress and the Executive Office, rather than the agency (civil service) level. While decision processes at the agency level are usually consistent and thorough, they become extremely ad hoc at the top levels of government, where technical questions are more likely to be viewed through the lens of political ideology than scientific metrics. Equally, the highest levels of government tend to be dominated by participants who may represent a broader array of interests than those characterised by the agency mission. In general, there is a lack of formal, objective, and logical processes devoted to the issues at hand.

The solicitation and use of outside scientific advice is generally poor at this level. Here, the advisor is often at arms length from the advisee, who frequently lacks the time to iterate conclusions and recommendations. When advice is provided by a panel or commission, it is more likely that stakeholders will play a dominant role than is the case at the agency level. The volumes of advice prepared for and most often funded by government are not well aggregated and understood – and not often implemented.

Policy makers are typically much more comfortable with the political or economic side of these issues. Moreover, in keeping with the scientific principle that all wisdom is subject to disproof, scientists express their conclusions in terms of theory rather than fact, and arguments are often seen by policy makers as pitting scientific "theory" against political or economic "truth". Thus, S&T policies are often politicised, and solutions derived are fractional, and narrowly focused. In fact, too often there is no "scientist at the table," to take part in these discussions. The U.S. Congress exacerbated this situation by eliminating their Office of Technology Assessment (OTA) in 1995. The lack of scientific knowledge among our political leadership is widely recognised. Representative Rush D. Holt, a physicist serving in Congress, has estimated that less than five percent of Members of Congress have any scientific or technical training [3]. While all decision processes must fully consider moral, ethical, social, and political aspects, technical content seldom gets the attention it needs.

3. Some Case Studies

It is instructive to consider just a few instances where the upper level of government has made poor decisions because of lack of attention to technical input. Reasons for their failures certainly include non-scientific factors, such as economic aspects. However, the issues are largely technical and good technical forecasting and analysis could have made significant differences to the outcome if that analysis had been heeded. Threats facing us dictate the necessity of minimising these mistakes. As science joins both sides of the war on terrorism, we must maintain the technological advantage and that demands good decisions and policies.

3.1 Superconductivity Supercollider (SSC)

Support for the SSC began at the Department of Energy in 1983, with a feasibility study. Following several stages of development, a Presidential decision to proceed with the SSC was made in early 1987 and potential sites were examined. Planned and approved for $8.5 billion by the government, and competed throughout the U.S., this research facility ended up under construction in Texas. The Texas National Research Laboratory Commission, formed in 1987 to oversee the Texas interest in the SSC, distributed approximately $100 million over a period of ten years to universities for SSC-related R&D.

The SSC itself was to consist of two 53-mile-long stacked rings. Beams of protons, cooled to near-absolute zero, were to be accelerated in opposite directions around the rings to nearly the speed of light by immense magnets. Their collision was to reveal secrets about fundamental particles. Construction was to have taken eight years, but, despite pleas from a host of notable scientists, the project was terminated by Congress in 1995, leaving the field to the Large Hadron Collider project proposed for CERN in Europe. The programme completed over 10 miles of track before calling it quits, with little to show for its investment.

The problem in this case study seems to have been one of commitment and prioritisation. There were no changes in S&T realities and motivation between SSC's start and cancellation. Cost estimates increased as the programme progressed, but no more dramatically than many other federal efforts that were successfully completed.

3.2 International Space Station (ISS)

The ISS was born from an effort in 1993 to reduce costs of an earlier plan called Space Station Freedom. The targeted annual expenditure was $2.1 billion. At first, the ISS was touted as an orbiting laboratory, where interesting and useful scientific research was to be accomplished. However, the space station was redesigned on several occasions to accommodate cost overruns. Each successive new design reduced the scientific capability of the station; the most notable was the loss of the centrifuge. Despite warnings from scientific advisors, such as the National Research Council, that the great cost of the ISS could not be justified by the value of the science to be done in orbit, the programme continued to be justified in that vein. In particular, the ISS was promoted as a laboratory for studying humans in space, even though the Russian Mir had been conducting these experiments for a decade or so. Emphasis shifted to its potential benefits for international diplomacy when it became clear that science alone could not justify the project. From the beginning Russia had been a partner, but that country's contributions diminished as its economic and infrastructure problems grew more massive. Other countries joined the programme but the U.S. ended up with a progressively larger share of the costs and overruns soared.

NASA made several moves toward commercialisation of the payload and then attempted to privatise the station itself. Unfortunately, industry was not interested in a "factory in the sky" that was accessible only episodically and at great expense. The symbiotic relationship between the ISS and the Space Shuttle was problematic in that the latter was an inefficient carrier. It has been much discussed since the Columbia accident that the shuttle programme, which was originally supposed to make 100 trips a year, made far less than that. In fact, there have barely been 100 trips during the twenty years of the programme. Finally, after two Shuttles were lost with crews, it became evident to all that the key to success in orbital space is not orbiting, but getting there and back, safely and affordably. All of these arguments were made along the way, but somehow Congress and a succession of Presidents failed to hear them and necessary shifts in investment to improve trans-atmospheric flight systems were never made.

3.3 National Flat Panel Display (NFPD) Strategy

Globalisation has had profound effects on U.S. competitiveness in producing and marketing products that utilise the technologies that Americans invent. Economic globalisation requires considerable attention from the S&T community when creating forecasts and development plans. An interesting example of an investment decision that did not work was the Department of Defense (DOD) flat panel display (FPD) strategy. In the early 1990s, the DOD began to express concern over the lack of domestic manufacturers in the business, a situation that meant that this important defence product would be available only from the overseas market. Many felt that dependency on others for such an essential weapon system component would hold the U.S. hostage during conflict. So under the National Flat Panel Display Initiative (NFPDI), the DOD invested millions of dollars to bolster the U.S. FPD manufacturing sector. However, studies in the mid-1990s revealed that the U.S. did not need to be concerned about foreign suppliers as it had in the past. In the early 1990s Japan was the only supplier of Active Matrix Liquid Crystal Displays (AMLCD) (the most common type of FPD), but by the mid 1990s several different countries had entered the market and there were over a dozen firms making FPDs. An Office of Technology Assessment (OTA) evaluation revealed that the supply was no longer in jeopardy. Finally, the AMLCD market was mature and operated on low profit margins that depended not on innovation, but high volume, improved manufacturing processes, and low wages. Given relatively high U.S. labour costs, Asia's superiority in manufacturing processes, and the U.S. ability to innovate, the OTA suggested that the U.S. focus on a next generation technology. While next generation technologies were pursued in the NFPDI, it was not done in a concerted manner. In fact, despite the overt objective to pursue AMLCD technology, the NFPDI did not get behind any particular technology in terms of actual resource allocation. Hence the manufacturing test bed sponsored by the NFPDI failed. Today, essentially no flat panel displays are produced in the U.S.

3.4 Human Genome

The Human Genome project must be counted as one of the most successful large U.S. scientific programmes. The goals were met below budget and earlier than scheduled. However, even though the project was originally begun to improve sharing of genetic information by developing it in the public domain, the project has produced a broad swathe of patents that include basic information and unique organisms. There is obviously a trade-off between the incentives to inventors and researchers represented by ownership of the intellectual property one creates, and the benefits of shared results of experimentation and analysis that might lead to new breakthroughs or better products. The proper balance has probably not been maintained, or even defined, and the major developer of publicly shared wisdom, academia, is now rivalling industry in its zeal to patent its findings. This shift from sharing to exploiting intellectual property has caused critics to question the tax-exempt status of some academic institutions. This may be a marginal issue in terms of national security, but it is crucial to U.S. global S&T competitiveness.

Our neglect of energy and public health policies will be discussed later. Suffice to say that the absence of consistency and good sense in planning for these two major national domains have blunted our future in a world that is dependent on both.

4. Levels of Government

It is worth a brief discussion of the key players in S&T policy and decisions at the highest level of government in the U.S. The two principal groups that participate in decisions at this level of government are the Executive and Legislative Branches. Although the Judicial Branch is sometimes part of the process, they seldom enter into decisions on S&T, unless they are called upon to adjudicate on the legal aspects of the strategies and policies. The one area where their role is very important is that of intellectual property rights where the Judicial Branch may have a considerable influence on S&T. For instance, patents being granted or renewed for modified pharmaceuticals and for organisms derived from genetic manipulation are being challenged on a legal basis. This represents a trade-off between placing important data in the public domain where everyone can use it and keeping it in the private domain to provide commercial motivation.

4.1 Executive Branch

The primary advisor to the President on S&T is the Director of the Office of Science and Technology Policy (OSTP). The mission of that Office, stipulated in the National Science and Technology Policy, Organization, and Priorities Act of 1976 (Public Law 94-282), is to serve as a source of S&T analysis and judgement for the President with respect to major policies, plans, and programmes of the Federal Government. The Act assigns several crucially important responsibilities to OSTP, listed below. The problem is that the OSTP is not able to fulfil these responsibilities.

- Advise the President and others within the Executive Office of the President on the impacts of science and technology on domestic and international affairs;
- Lead an interagency effort to develop and implement sound science and technology policies and budgets;

- Work with the private sector to ensure Federal investments in science and technology contribute to economic prosperity, environmental quality, and national security;
- Build strong partnerships among Federal, State, and local governments, other countries, and the scientific community;
- Evaluate the scale, quality, and effectiveness of the Federal effort in science and technology.

The Director of OSTP co-chairs the President's Council of Advisors on Science and Technology (PCAST), which draws from the private sector and academic community to provide advice on technology, scientific research priorities, and mathematics and science education. OSTP also supports the President's National Science and Technology Council (NSTC). The mission of the NSTC is to prepare coordinated R&D strategies and budget recommendations to orientate science and technology toward achieving national goals, which involves coordinating the various parts of the Federal R&D enterprise. To accomplish this, the NSTC reaches positions or coordinates actions in selected areas of S&T by assigning issues for study to Interagency Working Groups (IWG) formed by the Council and comprised of personnel from appropriate agencies [4]. The NSTC was formed in 1993 as "a virtual agency for science and technology" that would coordinate R&D across the government "to form a comprehensive investment package" that supports national goals.[1]

Congress created the Science and Technology Policy Institute, a federally funded R&D centre sponsored by the National Science Foundation. "Chartered by an act of Congress in 1991, STPI provides the highest quality and rigorously objective technical analytical support for the Office of Science and Technology Policy and other government users, under the sponsorship of the National Science Foundation".[2]

4.2 Legislative Branch

The elimination of the Office of Technology Assessment substantially increased the role of the Congressional Research Service (CRS) as a source of scientific advice to Congress. The CRS is the public policy research arm of the United States Congress. As a legislative branch agency within the Library of Congress, CRS works exclusively and directly for Members of Congress, their Committees and staff on a confidential, non-partisan basis. Under different names, CRS has advised Congress since 1914. Amongst the staff of the CRS are experts in natural sciences. Staff members work with Members of Congress and their staffs to produce written reports on numerous S&T issues. On specific issues the Congressional Budget Office (CBO) and Government Accounting Office (GAO) may also provide perspectives that affect or are affected by S&T. Common to CRS, CBO, and GAO is that their participation is by invitation only. A member of Congress must initiate an examination by one of these organisations, so their part in the process is seldom proactive and independent.

5. Two Types of Decisions

The Potomac Institute study considered two types of S&T policy issues that must be handled especially well: issues related to research investment and management ("policy-for-science"), and issues related to national security that have significant scientific content ("science-for-policy"). Because they are subject to so many pressures, participants, and agendas, the processes employed at the national level for making both science-for-policy and policy-for-science decisions vary greatly and tend to be rather ad hoc. This makes any attempt at a consistent description of how government works these issues difficult.

5.1 Policy for Science Issues

Government must invest in its S&T portfolio to provide technical solutions to society's problems (countering terrorism, for example). In the process of developing this portfolio, balance must be sought across a large spectrum of parameters, such as research versus applied research versus development, funding work in one discipline over another, directed versus non-directed research, advanced versus off the shelf technologies, and domestic versus foreign sources (the import/export balance). Other vital areas of concern are how to manage national laboratory systems to maximise their benefits, while reducing their considerable cost or how to ensure the adequacy of our technical workforce.

How can we ensure that decision makers allocate R&D resources and create S&T strategies that maintain an appropriate and consistent pressure on threats based on broad and objective assessments of national security, rather than swinging from one "threat du jour" to another? The answer must be found largely in doing a better job of forecasting and understanding all national security issues and in selecting long-term priorities that can best be confronted through the advancement and application of S&T. These are major challenges that are well beyond the scope of this Chapter, which addresses itself to what one does about formulating and maintaining a suitable R&D investment portfolio after these determinations are made.

From a procedural perspective, decisions on R&D investments follow a path from agency-to-President-to-Office of Management and Budget-to-Congress as shown in Figures 1 and 2. The agencies respond to Presidential guidelines, sometimes with coordination by the NSTC and the Office of Management and Budget (OMB), by developing the budgets that are passed to OMB in thirteen components.

To create a national R&D strategy, the NSTC develops research priorities and, together with the OMB, issues directives to agencies. It also issues directives on numerous budget issues, but it has no power over the budget after these directives are issued. Agencies submit their budgets back to the OMB, which considers the thirteen budgets separately. Before passing the budget onto Congress, the OMB aggregates the R&D budget, but does not examine it for consistency with national goals.

As shown in Figure 1, upon entering Congress, the House Science Authorization Committee considers a large portion of the R&D budget, but not DOD, and in reality their authorised levels matter little, since R&D programme decisions are then split up into the various appropriations subcommittees. There is virtually no coordination among the subcommittees regarding how separate R&D programmes may affect one another. The break up of the appropriations subcommittees mirrors the way that the thirteen appropriations bills are considered, so there is no opportunity for legislators to re-examine the collective R&D budget before separate sections are voted on.

As mentioned earlier, an NSTC role is to coordinate interagency research, a task that is sorely needed when science is increasingly interdisciplinary and we operate primarily through mission agencies, which frequently rely on the research of other departments. However, this organisation has failed to produce a cogent national R&D strategy to shape the U.S. R&D budget. The NSTC's 2000 Annual Report laid out the Administration's R&D priorities, but there was no discernable impact on the budget. While the goals of increasing funds to math and computer science and engineering were achieved, funding to environmental research and other fields of emphasis remained unchanged or declined. Moreover, though the importance of biological research was explained, there was no clear explanation of why it should receive more funding than all other non-defence funding combined, yet funding to the biological sciences continued to increase dramatically. Interestingly, in 2002 OSTP's President's Council of Advisors on Science and Technology (PCAST) issued a letter to the President recommending an increase in funding of physical sciences research, partly in response to the increase in funding for biological sciences [6].

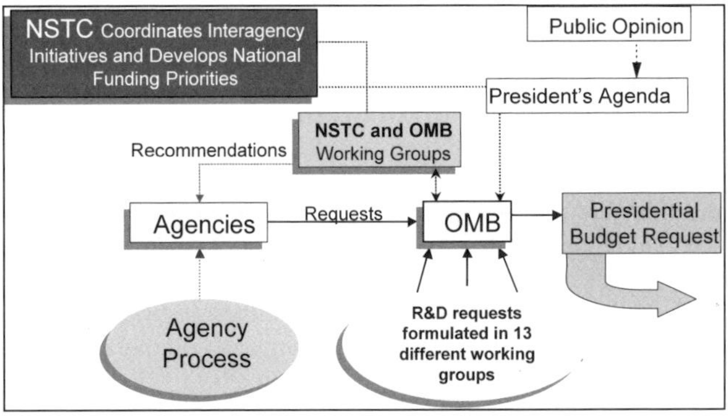

Figure 1. Agency Budget Preparation

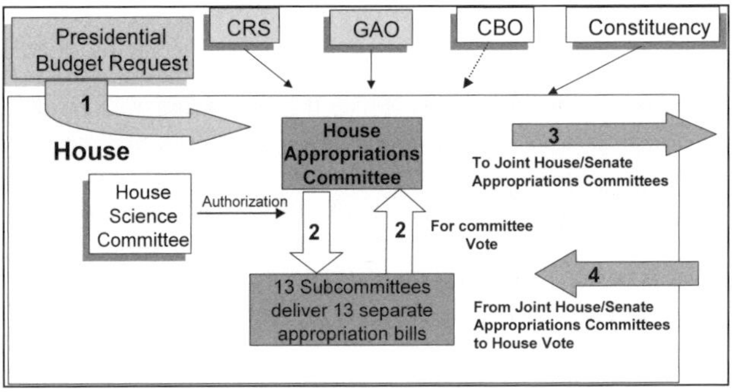

Figure 2. Presidential Budget to Congress to Appropriation

From its current position, the NSTC cannot develop and implement a national innovation strategy. The body has limited manpower, continuity, and purview to direct agencies toward the administration's priority areas, and once the budget has been prepared for Congress, it loses all control over funding distribution. Never again is the R&D budget considered in a truly holistic manner that promotes consideration of all programmes in relation to one another.

In keeping with its more centralised decision-making approach, the current administration may be taking steps to strengthen OMB's role [7]. The OMB and OSTP issued a joint memorandum in June of 2003, which contained a directive that not only affects interagency funding, but also exerts more influence over agency budgets. The memo includes updates to the Process Management Architecture (PMA) metrics, which include measures of programme relevance to national priorities. However, the main focus is to ensure that R&D is well managed, not to assess the relationship of the research to national goals.

While performance measurements existed in earlier memos, including the 2000 Annual Report, this metric is far more detailed. There is an indication that OMB will more carefully scrutinise research programmes for their contribution to national research priorities and research efforts. The PMA may indeed give the White House a way to closely monitor programmes and assess them for their benefit to national goals. If this were successfully accomplished, it would represent a powerful tool for science policy. The next step would be to carefully evaluate the aggregated R&D budget for consistency and gaps before it left the White House. In its 1995 report, chaired by Frank Press, the National Academies suggested that a committee be established to observe the aggregated R&D budget as it moved through the authorisations and, particularly, appropriations. This may help, but it would be difficult to evaluate the budget if it is still voted on separately. The key questions are whether the NSTC and OMB are the right organisations in the right place to do this and whether they can extend their influence to Congress. For reasons discussed later, the answer to both questions are no.

The implications are clear for the S&T efforts of the U.S. Department of Homeland Security (DHS). R&D budgets are finite and the nation must meet its national security S&T goals with as much efficiency as possible. The processes described above are anything but efficient. The DHS S&T budget is approximately $1 billion per year. Clearly, much more is needed and only through better overarching planning and collaborative interagency work can the Department reach its goals.

5.2 Science for Policy Issues

In the second decision category, the growing importance of science-laden national security issues emerges from three enduring trends. First, the increasing rate of breakthroughs in science. Second, the decreasing time between those breakthroughs and their widespread application. Third, the growing impact that these products of science can have on our lives and on national security. As part of the Potomac Institute study, we analysed each of these trends and found persuasive evidence that each is robust and enduring [1]. This means that national security issues with technical content will grow in number, complexity, and importance. Just a few examples of these issues are mentioned below.

5.2.1 Biological Sciences

We all believe that advances in biological sciences will produce untold benefits. For example, genetic research may lead to the discovery of cures for inheritable diseases such as schizophrenia and improved understanding of the body and brain and their functions are likely to yield remarkable new diagnostic and treatment techniques, as well as revolutionising how we learn. Enlightened policy is needed to ensure the continued flow of these blessings.

Yet there are also down-sides that must be dealt with through wise policies and decisions, such as the increase of drug-resistance, the appearance of new and virulent diseases, and wider economic gaps in health care between have and have-not patients. The creation of new species may endanger those already here, whether it is a genetically modified animal, pathogen, or pesticide-resistant crop. Our rapidly increasing knowledge of the brain and its functions will make physiological and psychological changes to humans temporarily possible through neurological implants and pharmaceuticals. This may provide a “dress rehearsal” for how to handle more permanent changes to be gained through genetic engineering.

The challenge of shifting the advantage to the defensive side of biological warfare and terrorism during the next twenty years has large investment policy ramifications. No one can discount the potential of tailoring biological weapons for greater virulence, longer life, or with a preference for a particular ethnic group – a frightening quiver of arrows in the hands of a rogue state or terrorist group.

5.2.2 Nanotechnology

While many applications of nanotechnology are speculative, we must marvel at their breadth and profound implications. If only they work! Academia and industry are already producing nano-scale products, such as coatings, microstructure fabrics, and fire retardants but there are also potential dangers, such as health and environmental risks in ingesting or inhaling nano-particles. These are harbingers of issues that the U.S. and the world will have to face if the benefits of nanotechnology are to be realised. We need to begin to develop policies that might be required to address the potential of these research products to go awry, but even more importantly, we must anticipate the likelihood that terrorists may use both the nano-scale devices and materials made available. Nano-scale toxicity or even the threat of a self-replicating "grey goo" may some day be tools of super-terrorism.

5.2.3 Energy

Over the next two decades we must meet the challenge of finding technologies that offer sufficient, affordable, and environmentally viable energy production. These technologies could yield major benefits to our national security, but this is not likely to happen unless energy R&D investments increase dramatically and a strong and targeted S&T programme emerges to focus our efforts on clean, renewable energy sources. That this is not happening in the United States is obvious from Figure 3, which shows a weak investment trend for energy research, compared with a nearly exponential climb for health research. Part of this policy must be a reasonable plan to protect our energy, electronic and grid-based infrastructure from terrorists – most particularly nuclear power plants. First, however, we need objective and effective energy goals and policies.

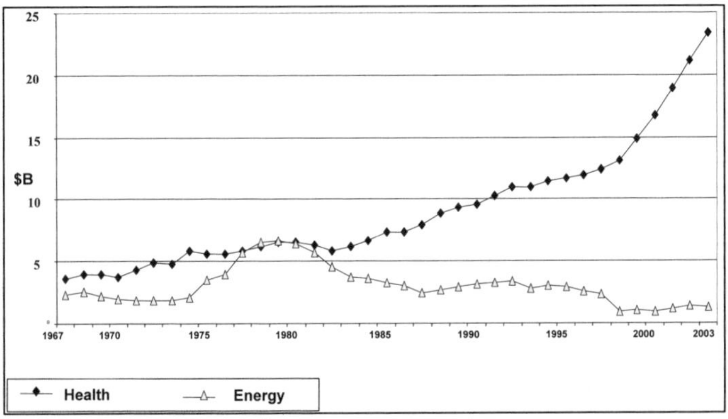

Figure 3. U.S. energy versus health research budgets (Source: [9])

5.2.4 Social, Behavioural, and Economic (SBE) Sciences

Two important national security issues involve the shortcomings in technical understanding of the general public to improve their judgment on S&T issues (voting on issues that involve science, for example) and the competitiveness of our technical workforce. Public interest in, and knowledge of, science and technology has been the subject of many polls and studies. For example, a 2001 NSF poll revealed that 47 percent of those questioned said they were very interested in new science discoveries. Medical discoveries were of particular interest, with 67 percent of respondents indicating they were "very interested." Only local school issues, which were placed second to medical discoveries, scored at the "very interested" level. This is the source of much of the pressure to increase health budgets, with 65 percent of those surveyed suggesting that they supported doubling spending on medical research during the next five years [9].

The way in which a society, particularly a democratic one, views science has a dramatic effect on the way science develops. How the various tensions are played out in the political process is reflected, at least schematically, in Figure 4, which illustrates that social forces can act to lubricate or create massive friction to dampen science investment.

Considering the profound relationship between society's perception of science and both policy-for-science and science-for-policy, it is critical that the American educational system improve in the areas of science and mathematics. The decisions of policymakers will not reflect the appropriate degree of scientific acumen, unless an educated public is willing and capable of holding them accountable of making poor decisions based on scientific merits.

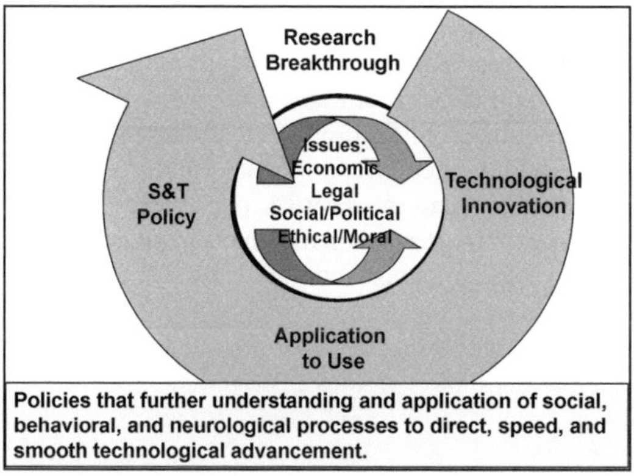

Figure 4. Forces of SBE on progress and application of S&T

6. Conclusions and Recommendations

An examination of these issues reveals that ideology and short-term fixes are too often substituted for depth and objectivity. An analytical element at the highest level of government is needed to focus attention on the long-term development of science policy. This idea is discussed briefly below, along with five other suggestions.

6.1 Develop and Codify a Sufficiently Broad Definition of National Security

There is little consistency in the definition of the term "national security". Perhaps a fairly broad definition should consider war, limited military operations, terrorism, natural threats, crime, threats to basic national principles, and threats to our economy and competitiveness. This scope may not be optimum, but it is a start on the path to finding and instantiating one that is, for while there has been great debate over what should be included in the definition, the current scope of U.S. national security policy seems inadequate to deal with the myriad threats that now confront our nation. With zero sum budgets, we will need synergistic defence against the many expected threats – a "Great Wall," rather than a series of partitions erected separately as each new (or old) threat enters centre stage. We need to encourage collaborative solutions among government and non-government participants if we are to seek common mechanisms to deal with problems with similar solution spaces, such as crime and terrorism, or biological agents and naturally occurring pathogens.

Other countries have had to deal with an evolving sense of what national security stands for. In discussions with S&T leaders of other countries, we found that the breadth of their national security definition depended almost entirely on what they faced in the near future. Some countries, such as South Korea, identified their security with the status of belligerent neighbours, while South Africa considered HIV/AIDS a primary national security threat. The more secure developing countries had the broadest definitions.

6.2 Establish a Separate Organisation for Science Policy

It may be time to establish an Office, Department, or Commission of Science Policy (OSP). Although this discussion has been tailored to the U.S, the basic argument would apply to some degree to other countries. The four major duties of the OSP would be forecasting national security issues that will emerge from important S&T trends, formulating the scientific context of these issues, preparing recommended policies to confront the issues, and following up on their implementation.

As discussed earlier, there are already U.S. organisations chartered to perform these functions, but unfortunately, they are not equipped to accomplish them because they have too few people or no budgetary voice, or they are in the wrong place organisationally. To be effective, the OSP cannot be a creature of either Congress or the President. The office must maintain sufficient continuity through successive administrations to resist the "see-saw effect" on issues battered by the move from one government to another. Such continuity, even if imperfect, is only found in an established civil service department or agency, which this smaller office would emulate. Such an organisation can better stick to a specific charter, rather than becoming sidetracked by subsidiary issues or temporary assignments.

One organisation, OSTP's National Science and Technology Council, could serve as a structural template for the OSP. Under our approach, some missions currently assigned to the NSTC could be shifted to the OSP, while others could either be assumed by the OSP or continue to be handled by the NSTC.

Interestingly, it seems clear that the success of an OSP would depend to a great extent on carefully defining its jurisdiction. The OSP should concentrate on the tasks suggested above and only for national security issues (broadly defined). Its staff must be willing to seek and evaluate advice effectively and they must know the basics of science sufficiently to comprehend the advice given. They must also have the breadth of knowledge to introduce other aspects of the issues. They must understand and contribute to S&T policies and be able to work collaboratively with the staffs of other nations on S&T issues with international scope. Moreover, they must be sufficiently versed in the policy and political world to get the best advice implemented.

There is a history here. The first serious attempt at a Department of Science was during the 1880s, when the Allison Commission put forth the idea of bringing together all scientific offices and bureaus under one national department. However, the laissez faire spirit of the era prevailed, with Congressman Hilary Herbert noting, "Government patronage shackles that spirit of independent thought which is the life of science."

More recently, in the period after Vannevar Bush's report to President Roosevelt various national scientific offices were established including the National Science Foundation, NASA, Office of Naval Research and the Atomic Energy Commission. There have been 60 episodic attempts in Congress to consolidate the various Federal science and technology agencies. During the Reagan Administration, the President's Commission on Industrial Competitiveness proposed consolidating Federal S&T initiatives under a new Department of Science and Technology. Bills to establish a Department of Science and Technology (or some similar designation) continued into the mid-1990s.

Historical arguments for a Department of Science have included improving: R&D budget planning and balancing, uniformity of federal science policy implementation, and the effectiveness of intellectual cross-fertilisation. On the other hand, arguments against cite a potential to: stifle diversification of federal S&T capabilities that have proven to be effective and helpful to national welfare, specific agency mission impacts due to centralisation, reduce individual initiative and promote dependence on government [10, 11].

6.3 Increase the Input of S&T Agencies to the Highest Level of Government

Clearly, we need to do a better job of attracting S&T agency skills into the process of developing S&T policy. The bulk of government technical acumen lies in those agencies and although many scientists make it to top positions in government it seldom occurs while they are still actively engaged in research. Perhaps we could encourage a kind of "intellectual tithing," to bring skills to bear; beginning at the top of agencies with an S&T Council composed of the directors of government S&T agencies (similar to what was originally proposed to aid the Department of Homeland Security to initiate its S&T programme). This approach would also bring together the best and brightest in each field of concern, whether in academia, government, or the private sector. This Council would develop positions on each field in question, providing insights and hard facts on trends, timelines, opportunities, and threats.

6.4 Improve the Process of Seeking, Evaluating, Aggregating, and Implementing S&T Advice at the Highest Level

There is ample S&T advice available to the government, but there is often no one to aggregate and then iterate or build upon emerging recommendations so that appropriate policy and decisions result. Government seldom exerts the efforts necessary to seek and handle external advice effectively. There is a large body of literature on ways to make this process better and we discuss some of them in our Potomac Institute report [1]. For example, there are difficulties that arise from society's reliance on researchers themselves to evaluate the risk in their fields of endeavour. Most professional organisations that represent areas of technology are advocates, and understandably would rather dwell on the positive potential of their research than on its dangers. Of course, there have been exceptions, such as Bill Joy at the Foresight Institute, and, on occasion, the National Science Foundation.

Regardless of how the processes of seeking advice are improved, the new Department of Homeland Security should be among the first to do so. Despite the "everyone under the same umbrella" claims that accompanied the initiation of DHS, no one organisation has the immense expertise that anti- and counter- terrorism demands. To do its job well, the DHS needs all the help it can get.

6.5 Promote National Security S&T Policy Offices in the Executive Branch and Congress

Whether or not the idea of an OSP is rejected, there is another organisational device that would be helpful in developing accurate and consistent technical foundations for issues shared between the President and Congress. Under this concept, two organisations, one in the Executive Branch and one in the Legislative Branch, would be dedicated to developing and documenting a technical understanding of important issues with significant S&T content and forecasting S&T impacts on society. Unlike the OSTP, Congressional Research Service, or the late Office of Technology Assessment, these organisations would be proactive and have a long-term focus in accomplishing this mission. They would forecast S&T issue areas that are likely to require attention at the highest level of government and set the stage for decision-making. Ultimately, the two organisations would meet and develop a common technical picture of the scientific aspects of the issue (or at least to know where their two pictures differ).

6.6 Improve Communication with Political Decision-Makers through the Use of Technology

Emerging technologies to explain difficult S&T concepts behind issues and to meld technical and non-technical aspects are becoming available. They should be adopted in discussions with top government decision makers.

These recommendations should be just one volley in an interactive discussion of how government faces a future of vitally important S&T decisions. Ignoring the realities and practicalities of the political system would be disastrous to the task of implementing the proposed actions, but likewise nations should consider the need for new, even extra-political, perspectives in developing better ways to shape S&T to serve national security.

Notes

1 See the Office of Science and Technology Policy Website: http://www.ostp.gov/NSTC/html/execorder12881.html.

2 See the Institute for Defense Analyses Website: http://www.ida.org/stpi/index.html.

References

[1] Richardson JJ, Matson WB, Peters RJ. Innovating science policy: restructuring S&T policy for the twenty-first century. Review of Policy Research 2004; 21 (6): 809-828.

[2] Dealing with democracy. Nature 2003; 425, 25 September: 329.

[3] Holt RD. Bringing science and technology back to Congress. American Psychological Society Observer April 2002. Available at: http://www.psychologicalscience.org/observer/0402/holt.html.

[4] National Science and Technology Council (US) [NSTC] 2000 Annual Report. Washington DC: NSTC; 2001. Available at: http://www.ostp.gov/NSTC/html/nstc_ar.pdf.

[5] Congressional Research Service (US) [CRS] Homeland Security and Counterterrorism Research and Development: Funding, Organization, and Oversight. Washington DC: CRS; updated May 19 2003.

[6] President's Council of Advisors on Science and Technology (US) [PCAST]. Letter to the President dated 16 October 2002. Available at: http://www.ostp.gov/NSTC/html/execorder12881.html.

[7] OMB Watch. OMB expands influence over scientific decisions. Washington DC: OMB Watch; May 28, 2003.

[8] Allocating Federal Funds for Science and Technology. Committee on Criteria for Federal Support of Research and Development. Washington DC: National Academies Press; 1995.

[9] National Science Board (US). Science and Engineering Indicators – 2002. Arlington (VA): National Science Foundation; 2002.

[10] Dupree AH. Science in the Federal Government: a History of Policy and Activities. Baltimore: The Johns Hopkins University Press; 1986.

[11] Boesman WC. A Department of Science and Technology: A Recurring Theme. CRS Report for Congress 95-235. Washington DC: Congressional Research Service; February 3 1995.

Science and Technology Policies for the Anti-Terrorism Era
Andrew D. James (Ed.)
IOS Press, 2006

What Can the Science and Technology Community Contribute?

Bill DURODIÉ
Cranfield University, Defence Academy of the United Kingdom,
Shrivenham, Swindon, Wiltshire SN6 8LA, United Kingdom

Abstract. This Chapter explores the role attributed to science and technology in combating the global war on terror in an age when social bonds have been eroded and our sense of the need for social solutions diminished accordingly. One consequence of this is the exaggeration of risks presented by science and by terrorists to the point of ignoring the more mundane and probable threats that confront us. Prioritising technical means to build social resilience over cultural change is also likely to be counter-productive by further fragmenting the ordinary human bonds that actually make society truly resilient. A political debate over societal values is required if we are to re-engage the public and deal appropriately with all-manner of disasters, including terrorist attacks.

1. Introduction

Science and engineering have always played a part in war. The advent of new technologies has only increased this potential role. The global war on terror is no different to other wars in that regard. Many proposed options for dealing with terrorism have an explicit technological angle. These include the need for better intelligence and surveillance, the development of new instruments for detecting chemical, biological and radiological agents, specialist clothing and equipment for emergency responders, and computer models for predicting behaviour or orchestrating responses.

It is understandable, even commendable, that well-meaning experts and professionals should wish to get involved. Further, a significant amount of social resources are being diverted to tackling the problems raised. Accordingly, those with an eye on sources of funding to explore new areas of inquiry are likely to be interested. Indeed, beyond the explicit development of technical capabilities, the war raises numerous implicit issues for scientists and engineers to deal with. Who has access to the technologies they develop? How much should be made available in the public domain?

Before diving in off the deep end, however, those of a more critical disposition – as any true scientist should be – would do well to examine the broader context within which these events and issues have occurred and how they have been framed. Things are rarely as they seem. The primary task of all concerned ought to be to grasp the underlying essence of what is going on. Failing to do so could lead to the development of proposed solutions that, at best, merely contain perceived threats, at worst, exacerbate them significantly, not least by undermining our own capabilities to be resilient in the long run.

Many perceived problems in the world today are driven more by their social context than by their scientific content. Scientists and engineers need to be alert to this, not least because science occupies a peculiar position in contemporary life. A diminished sense of the significance of, as well as the desire and ability to shape, social forces, has led to an increased focus on the importance and impact of science upon our lives. In response to this elevation and exaggeration of science, society has increasingly become preoccupied with science as a potential source of new risks.

This has led to the highlighting and fetishisation of purported scientific and technological solutions to what remain essentially social problems, as well as a concomitant and distorted perception of threat from anything remotely scientific in content. A recent publication from the Royal Society, the United Kingdom's leading scientific institution, is quite apposite in this regard [1]. The report; *Making the UK Safer: Detecting and Decontaminating Chemical and Biological Agents*, is undoubtedly rigorous in scope and methodology. However, it is the unquestioning acceptance of the social context that needs examining.

In it, some of the UK's leading scientists take at face value the notion that: "Recent global events have given greater prominence to the threat of chemical and biological agents being used malevolently against civil targets", and further that: "Science, engineering and technology are central to reducing this threat". Both of these assumptions would benefit from interrogation. Indeed, questioning the axioms of a debate ought to be the first step in making it truly objective. Otherwise we may be left with a technically competent, but ultimately unscientific report.

It is not just the job of social scientists, but scientists too, to question whether this purported "greater prominence" is real. Assuming that it is, scientists true to their tradition would then start by asking what this fact represents. Whether this is a media construct, or a more deeply held social concern, across different layers of society. If it is the latter, it ought to be considered that such a concern may have little relation to the actual probability of the threat they fear. The fact that something is possible may cause alarm, but is the best way of assuaging this to assume those fears to be real and then seek to mitigate their outcomes, or alternatively, to interrogate those fears?

Ultimately, the Royal Society report may be of use to a highly limited number of technical specialists who, in the extremely unlikely eventuality of such a situation arising, would be charged with dealing with it. However, it is not obvious what its use is beyond that, in the public domain. Surely, publication of the report itself could now serve to confirm people's exaggerated perceptions of threat? It has certainly contributed to the "greater prominence' that it originally sought to address. People might assume that if the UK's leading scientists are investigating such matters then their presumptions are more likely to be true.

2. Science and Society

The emphasis often given as to the importance of science for effecting social change is one-sided. Science can transform society, but it is also a product of society. Its advances and remit, as well as being shaped by material reality, are circumscribed by the nature and values of the society within which it develops. The ambition and imagination of that society – or lack of these – is important here. Hence, whilst the world of antiquity yielded many intellectual insights, constrained by its social structures, these proved to be of little practical consequence [2].

It was only when the largely static feudal order dissolved, through the development of trade, that new demands were raised on individuals and society. A marriage of intellectual activity with practical needs encouraged innovation and, through the accumulation of wealth, challenged the old social order. As well as delivering remarkable achievements, social and scientific developments raised expectations as to what was possible [3]. This was about more than simply an advance in scientific knowledge – it was part of a wider shift in attitudes and beliefs.

The aspiration for social progress gave humanity confidence in the power of its own reason – a factor that then proved of significant importance to the development of science. The Scientific Revolution represented the triumph of rationality and experimentation over the superstition, speculation, diktat and domination that had gone before. It was a practical battering-ram with which to challenge perception, prejudice and power. But science was merely the product of a broader social dynamism, as well as becoming an essential contributor to it.

Just as the initial dynamic behind science was social change, so social change, or more particularly the lack of it, could circumscribe it too. Initially this came from the vociferous rejection of the old religious and monarchical orders it had supplanted. Then the advent of positivism consciously sought to restore order by decoupling science from wider political aspirations to transform society [4]. This reflected the inherent limitations and world view of the new industrial elite who derived their wealth and influence from simple mechanical processes linking cause and effect by uniform rules.

However, over the course of the twentieth century a wider layer of society lost its faith in the progressive capabilities of scientific transformation. Two world wars, separated by a depression and followed by continuing poverty and conflict in the developing world generated doubts as to the possibility of universal human progress [5]. Radicals, who had traditionally championed the liberating potential of scientific advance, now came to view it with increased suspicion. They also associated the Manhattan project and the Apollo programme with American militarism.

Some now argued that aspiration itself – rather than its failure as evidenced in the collapse of confidence in social progress – was dangerous [6]. Science was seen as the amoral steamroller of a dispassionate new modernity that crushed communities and tradition. What is so poignant about the modern disenchantment with science, is that it has emerged at a time when its achievements are without precedent. Behind the current crisis of faith in science, however, lies a collapse of confidence in humanity, and hence in the desirability and possibility of social transformation [7].

The defeat of the old Left externally, symbolised by the disintegration of the former Soviet Union and its satellite states, and the taming of the Left internally, symbolised in the UK through a series of political defeats over the course of the 1980s, now led it to make new alliances, including with the environmental movement – traditionally the preserve of the romantic Right – in order to boost its numbers, and leading it to shape a new, more individual or consumer-oriented agenda. At the same time, the diminished sense of the possibility of shaping social factors also made science appear to play a more important role in determining things.

3. Social Erosion

In parallel with the gradual disillusionment of society with science, has come an equally significant process of disengagement of society from politics. For the vast majority of ordinary citizens this has been exacerbated by a growing sense of social disconnection. At both the formal and informal levels of social engagement, social bonds have been severely eroded over the last decade or so. The resultant sense of isolation and insecurity across society has become the key element shaping perceptions of risk.

At the formal level, people in advanced Western societies are increasingly unlikely to participate in the political process. This effect is most striking among younger age groups. Electoral turnouts are at an all-time low and in the few instances where these are high, emotion appears to rule over reason. Few are active, or even passive, members of political parties or trade unions as their forebears were, and there is little attempt to engage in, or raise the standard of, debate. When people do vote, it is often on a negative basis – against an incumbent, rather than for a replacement.

At the informal level, the changes are even more striking. Many have commented on the growing pressures faced by communities, neighbourhoods and families. In his book on this theme, "Bowling Alone", the American academic Robert Putnam also pointed to the demise of informal clubs and associations [8]. Meeting up with friends, occurs less frequently than previously too. In other words, people are not just politically disengaged but also, increasingly socially disconnected. This loss of social capital has occurred and been experienced within a generation.

Not so long ago, for example, it was still possible across most urban centres, to send children to school on their own, assuming that other adults would act *in loco parentis* – chastising them if they were misbehaving and helping them if they were in trouble. Today, such a straightforward social arrangement can no longer be taken for granted. None of us ever signed a contract saying that we would look after other people's children. It was simply an unstated and self-evident social good. This loss of a social sense of responsibility makes the individual task of parenting harder [9].

In a similar way, ordinary communities, at the turn of the last century, invested a great deal of effort in establishing and running their own institutions. These took a wide variety of forms from churches, to working men's clubs, schools and trade unions. It is almost impossible to find a similar process at work within society today. This is not to suggest some kind of golden-age of community activism. Clearly, past societies were also associated with a wide manner of activities we are quite glad to have seen the back of. However, the resulting erosion of social connectedness is significant.

Being less connected, leaves people less corrected. It allows their subjective impression of reality to go unmediated or unmoderated through membership of a wider group, association or trusted community. Without a sense of the possibility of social solutions, personal obsessions grow into all-consuming worldviews that are rarely open to reasoned interrogation or debate. In part, it is this that explains our recent proclivity to emphasise or exaggerate all of the so-called risks that are held to confront us [10].

Rather than the world changing any faster today than in the past, or becoming a more dangerous, unpredictable or complex place, it may be our diminished, and more isolated, sense of self that has altered our confidence to deal with change and the problems it gives rise to [11].

Those who talk of a "Runaway World" [12], would be hard pressed to show how the pace of change today is any greater than say, over the sixty-five year period two centuries ago between the creation of Richard Trevithick's first steam locomotive and the advent of transcontinental railroads across the United States of America. Alternatively, note the pace of change over the same period a century ago between the Wright brothers' first powered flight and man walking on the moon. If anything, change today appears somewhat attenuated.

Much of the focus recently has been on the largely undelivered promises of biotechnology – a technology now passed its fiftieth anniversary – and the potential of the internet. But whilst the latter may have led us to being more networked virtually, it has not driven much change in the real world. Radically overhauling existing transport networks, a transformation not currently envisaged, would most likely have greater social and scientific consequences.

In our technically networked world, we may be more aware – but we are also easier to scare – than previously. Being more isolated leaves us more self-centred, as well as risk averse. In turn, these developments reduce the likelihood of our acting for some greater common good and end up making us less resilient, both as individuals and as a society.

From BSE to GMOs; from mobile phones to MMR, all new developments are now viewed through the prism of a heightened and individuated consciousness of risk. Nor are our fears restricted to the realms of science and technology. Age-old activities and processes have been reinterpreted to fit our new sense of isolation and fear. Bullying, sun-bathing and even sex have joined an ever-growing panoply of concerns, along with maverick doctors, crime, food and paedophiles.

Worse, this state of affairs has been exacerbated by the various authorities themselves, who suffer from their own existential crisis of isolation and insecurity. As we no longer vote, so ruling parties appear increasingly illegitimate and divorced from everyday concerns. A less than 50% turnout when split two or three ways produces governments with at best a 20-25% mandate. The real figure as reflected by demographics, negative voting and actual local election results is often well below this, languishing around the 10-15% mark.

This crisis of legitimacy has been further accentuated by a certain lack of purpose that has set in since the dissolution of the old Cold War divide. Then, an ideological divide separated a supposedly socialist Left from the free-market Right. Far from the demise of the Left revealing the "End of History" [13], it actually exposed the Right's own lack of ideas and dynamism. In an age when social change has been problematised, the pursuit of profit through innovation no longer bestows moral authority as easily. Now all parties fight for the centre ground and desperately seek issues that mitigate change and will re-connect with voters.

Latching on to the general climate of fear and insecurity, politicians have learnt to repackage themselves as societal risk managers around issues such as security, health and the environment. They pose as the people who will protect us from our fears and regulate the world accordingly. But the petty lifestyle concerns they focus on, as reflected in incessant debates about smoking, smacking, eating and drinking are unlikely to inspire and engage a new generation of voters. Nor will doom-laden predictions relating to terrorism and global warming.

Indeed, the more such concerns are highlighted, the more it becomes impossible for the authorities to satiate the insecurities they create. Hence, alongside disengagement and alienation, has come a concomitant disillusionment and mistrust in all forms of authority, whether political, corporate, or scientific. Healthy scepticism has increasingly been replaced by unthinking cynicism. In many situations today, the public tend to assume the worst and presume a cover-up. Rumour and myth abound over evidence and reason.

4. Creating Fears

At a recent forum in London, a member of the security service informed an audience of bankers that, whist it was true that the probability of a chemical, biological, radiological and even nuclear terrorist attack was low, this could not be ruled out. It was suggested that groups such as Al Qa'ida may have relatively poor capabilities in such techniques but their intention to develop these was clear, and if they did the consequences might be devastating.

This, in essence, captures the logic of our times; "Never mind the evidence, just focus on the possibility". It is a logic that allows entirely vacuous statements such as that of an official after the supposed discovery of the chemical agent ricin at a flat in North London, who was reported as saying; "There is a very serious threat out there still that chemicals that have not been found may be used by people who have not yet been identified" [14].

Yet undiscovered threats from unidentified quarters have allowed an all-too-real reorganisation of everyday life. The U.S. government has provided $3 billion to enhance bioterrorism preparedness [15]. Developed nations across the globe have felt obliged to stockpile smallpox vaccines following a process, akin to knocking over a line of dominoes, whereby one speculative "What if?" type question, regarding the possibility of terrorists acquiring the virus, leads to others regarding their ability to deploy it, and so on. Health advisories to help GPs spot the early signs of tularemia and viral haemorrahagic fever have cascaded through the UK's urgent alert system. Homes across the land have received the government's considered message for such incidents; "Go in, stay in, tune in" [16].

Like all social fears, there is a rational kernel behind these concerns. Yet this is distorted by our contemporary cultural proclivity to assume the worst. It is the fear of bioterrorism that is truly contagious, and it is a fear that distracts us from more plausible sources of danger, diverting social resources accordingly, and exposing us all to greater risk [17]. It is also a fear that has bred a cynical industry of security advisors and consultants, out to make a fast buck by exploiting public concerns, and thereby driving those concerns still further.

There is a long history of bioterrorism incidents of which the anthrax attacks on politicians and the media in the U.S. in 2001 were but the latest [18]. Corpses infected with bubonic plague were thrown over the walls of Kaffa by the Black Sea in the mid-fourteenth century. At best, these are tactical devices with limited consequence, but not strategic weapons. It is the advent of biotechnology and the more recent, if overstated, possibility of genetically engineering agents to target biological systems at a molecular level, that is now held to pose a new challenge [19].

Few commentators point to the difficulties in developing, producing and deploying biological agents. This is evidenced by the failures of the Japanese cult, Aum Shinrikyo, in this regards only a decade ago. It was this that led them to settle for the rather more limited impact produced by the chemical agent sarin, despite their resources and scientific capabilities [20]. The Tokyo subway attack that ensued had rather more impact upon our fevered imagination, than in reality.

As with the anthrax attacks, this incident suggested that bioterrorism is more likely to originate amongst malcontents at home, due to greater access and capabilities in developing such weapons there. Advanced economies are also better placed to deal with the consequences of bioterrorism, a fact that significantly undermines their purpose, especially to outsiders. Nevertheless, suicidal foreign malefactors bent on undermining western democracies continue to be presented as the greater threat.

Recognising the extremely low probability and limited consequences of such incidents, some scientists point to the longer-term psychological impacts as being the more important [21]. There is an element of truth to this. Psychological casualties are a real phenomenon. In certain emergencies these can rapidly overwhelm existing healthcare resources and thereby undermine the treatment of those more directly affected [22]. Yet they can also become a self-fulfilling prophecy. Indeed, by increasingly framing social problems through the prism of individual emotions, people have been encouraged to feel powerless and ill [23].

The arrival of television cameras or emergency workers wearing decontamination suits act as powerful confirming triggers for the spread of mass psychogenic illness [24]. So too can psychosocial interventions, such as debriefing subsequent to an incident [25]. These can undermine constructive, pro-social and rational responses, including the expression of strong emotions such as anger [26]. Hence, despite good intentions, psychiatrists can become complicit in shaping social ills. This is because few are prepared to question the dominant cultural script emphasising social and individual vulnerability, and the need for professional intervention and support.

Rather than critically questioning the framing of the debate, many, like the scientists of the Royal Society mentioned earlier, now simply accept the possibility of chemical, biological, radiological and nuclear terrorism as a given. There is little understanding of how our exaggerated sense of risk is both historically contingent, predating 2001 quite significantly, and culturally determining, giving shape to and driving much of the agenda.

One medical historian and epidemiologist, has noted that "experts were using the threat of novel diseases" as a rationale for change long before any recent incident, and that contemporary responses draw on "a repertoire of metaphors, images and values" [27]. He suggests that "American concerns about global social change are refracted through the lens of infectious disease". This coincides with the view of others who see bioterrorism as providing a powerful metaphor for elite fears of social corrosion from within [28].

Despite incidents since 2001 pointing to the preferred use of car bombs, high explosives and poorly deployed surface-to-air missiles, the authorities have, through their pronouncements, encouraged the media to hype weapons of mass destruction. This is despite any terrorist's capabilities being pathetic compared to our own and the consequences being more likely to devastate them than us. We have stockpiled smallpox vaccines, but notably, have run out of influenza jabs. In the extremely unlikely eventuality of an incident occurring, we assume that the public will panic and be unable to cope without long-term therapeutic counselling.

In an age readily gripped by morbid fantasies and poisonous nightmares, few surpass the pathological projection of our own isolation much better than the fear of bioterrorism. All of this rather begs the question as to who is corrupting civilisation the most. The fantasy bombers or the worst-case speculators?

5. Cultural Responses

In fact, how we, as individuals and as a society, define and respond to disasters, is only partly dependent upon causal agents and scale. Historically evolving cultural attitudes and outlooks, as well as other social factors, play a far greater role. In objective terms, risk may be defined as a function of hazard and probability, but that some product or event is perceived of as a risk, or is treated as a disaster, depends on subjective factors.

This human element is missing from mechanistic risk calculus and technical solutions. Technical definitions of risk and resilience not only omit key elements of understanding and response - such as our degree of trust in authority, in other human beings and in ourselves – but may also serve to further undermine such factors, which are crucial in responding effectively.

The contemporary cultural proclivity to speculate wildly as to the likelihood of adverse events and to demand high-profile responses and capabilities based on worst-case scenarios may, in the end, only serve to distract attention and divert social resources in a way that is not warranted by a more pragmatic assessment and prioritisation of all of the risks that we face.

Technique and technology certainly help in the face of disaster. Ultimately, however, the fact that particular societies both choose and have the capacity to prioritise such elements, is also socially determined. More broadly, it is possible to say that resilience – loosely defined as the ability of individuals and society to keep going after a shock – is most definitely a function of cultural attitude or outlook. It is not an item that can readily be purchased.

Cultural values point to why it is that, at certain times and in certain societies, a widespread loss of life fails to be a point of discussion, whilst at other times or in a different society, even a very limited loss can become a key cultural reference point. This evolving context and framework of cultural meanings explains such variations as our widespread indifference to the daily loss of life upon our roads, as opposed to, for instance, the shock and national mourning that ensued from the loss of just seven lives aboard the Challenger spacecraft in 1986.

The loss of Challenger represented a low-point in our cultural assessment of our own technological capabilities. It was a blow to our assumption of steady scientific and technological progress that no number of everyday car accidents could replicate. It fed into and drove a debate that continues to this day regarding our relationship with nature and a presumed human arrogance in seeking to pursue goals beyond ourselves.

Hence, emergencies and disasters, including terrorist attacks, take on a different role dependent upon what they represent to particular societies at particular times, rather than solely on the basis of objective indicators, such as real costs and lives lost. In this sense, our response to terrorist incidents, such as that which occurred on September 11th 2001, teaches us far more about ourselves than about the terrorists [29].

On the whole, the history of human responses to disaster, including terrorist attacks, is quite heartening. People tend to be at their most co-operative and focused at such times. There are very few instances of panic [30]. The recent earthquake and tsunami in the Indian Ocean serve as a salutary reminder of this. Amidst the tales of devastation and woe, numerous individual and collective acts of bravery and sacrifice stand out, reminding us of the ordinary courage and conviction that are part of the human condition.

People often come together in an emergency in new, and largely unexpected ways, re-affirming core social bonds and their common humanity. Research reveals communities that were considered to be better off through having had to cope with adversity or a crisis [31]. Rather than being psychologically scarred, it appears equally possible to emerge enhanced. In other words, whilst a disaster, including a terrorist attack, destroys physical and economic capital, it has the potential to serve as a rare opportunity in contemporary society to build-up social capital.

Of course, terrorists hope that their acts will lead to a breakdown in social cohesion. Whether this is so, is up to us. Civilians are the true first responders and first line of defence at such times. Their support prior to, and their reactions subsequent to any incident, are crucial. Disasters act as one of the best indicators of the strength of pre-existing social bonds across a community. Societies that are together, pull together – those that are apart, are more likely to fall apart.

Whilst there is much empirical evidence pointing to the positive elements of ordinary human responses to disaster, it is usually after the immediate danger has subsided that the real values of society as a whole come to the fore. It is then that the cultural outlook and impact of social leaders and their responses begins to hold sway. These determine whether the focus is on reconstruction and the future, or on retribution and the past. A more recent development has been the trend to encourage mass outpourings of public grief, minutes of silence or some other symbols of "conspicuous compassion".

Sadly, despite the variety of ways in which it is possible to interpret and respond to different emergencies, the onus today seems to veer away from a celebration of human spirit and societal resilience, towards a focus on compensation and individual vulnerability. In large part this is driven by a narrowly technical view of risk and resilience.

6. Technical Resilience

Since September 11th 2001 much focus has been placed upon the concept of resilience, understood as the ability to withstand or recover from adverse conditions or disruptive challenges. Politicians, emergency planners and others, talk of the need to "build", "engender", "improve" or "enhance" resilience in society [32].

Unfortunately, much of this debate is framed in the fashionable, but limited, language of risk management and risk communication. Senior officials regularly point to the central role they attribute to risk reduction. This is understood in narrowly technical terms as consisting of horizon scanning, investment in equipment, training, business continuity planning, new legislation and the like [33].

This outlook itself reveals the absence of purpose and direction in society at large. After all, risk reduction is a means, not an end. In the past, society was not so much focused on reducing risk as upon enhancing capabilities towards some wider goal. Risk reduction was a by-product of such broader purposes and activities.

Presumably, people were prepared to risk their lives fighting fires or fighting a war, not so that their children could, in their turn, grow up to fight fires and fight wars, but because they believed that there was something more important to life worth fighting for. It is the catastrophic absence of any discussion as to what that something more important is, that leaves us fundamentally unarmed in the face of adversity today. In that regards, risk management is both insufficient as an approach, as well as being fundamentally unambitious.

It is also worth noting, that in recent times, the concept of risk itself has gradually altered from one that captured possibility and engagement in the active sense of "taking a risk", to one that increasingly reflects our growing sense of doom and distance, as evidenced in growing reference to the passive phrase of "being at risk". Risk used to be a verb. Now it has become a noun.

This is a reflection of the wider passive disengagement across society at large and further drives this by gradually removing our sense of will and agency from the equation. Risks are now conceived as being entities in their own right, only minimally subject to human intervention [34]. They are inherently and implacably out there, coming our way. The best we can do is to identify them and prepare to deal with them.

Even when discussing prevention, the assumption is that we are merely anticipating and building capacity for "inevitable" challenges [35]. In the words of some senior officials, it is "only a matter of time", or "when, not if", a terrorist atrocity will occur in the United Kingdom using some kind of crude chemical, biological or radiological device [36]. The notion that it may be possible to shape conditions, or set the agenda, with a view to obtaining more desirable outcomes or altering our social mindset, independently of external forces, is rarely entertained.

Unfortunately, much of the rhetoric regarding the war on terror, far from being robust and resolute, reveals an almost resigned fatalism towards future events. There is no sense of changing *how* people will respond, simply a sense of preparing them *to* respond. This defensive responsiveness in turn can only further encourage, not just terrorists, but a whole host of other malcontents, loners, hoaxers and cranks in their activities.

At best, our strategy is one of reacting to the presumed actions of others. They drive – we follow, or mitigate. Despite occasional references to the need to "defend our way of life" or "our values" [37], very little effort has been put into identifying what these might be. They tend to be assumed, or glossed over, in some cursory fashion. At best, tolerance, which is the passive virtue of putting up with other people's values, gets misconstrued as an active value.

No doubt, because societal aims and cultural values are deeply contested and debating these might appear to be divisive at a time when we need to act in unison, it is easier to face the other way. Yet this flagrant lack of clarification as to who we are, what we believe in and where we are heading as a society fundamentally undermines any technical attempt to be resilient.

Real resilience, at a deeper social level, depends upon identifying what we are for, not just what we are against. That way we can orientate society and seek to build upon it, not just anticipate what is coming and seek to respond. It is precisely by establishing our aims and values and then pursuing these, that we stand the most chance of winning hearts and minds, not just at home but also amongst the disaffected abroad.

This is not to deny the need for a small layer of highly-trained professionals in society to deal with the problem of terrorism in the here-and-now. Yet the debate about who we are and what we are for is not some abstract philosophical issue waiting for present hostilities to be over. It is most urgent and necessary right now. Without an eye on the ends, just as much as on the means, we may take decisions that drive us further from our goals than we appreciate.

What we do in the present, including the science and technology we develop, is inevitably shaped by our existing values, as well as the form of society we seek to create. There are already many signs that some of the actions that have been taken thus far have served to further exacerbate the deep mistrust and cynicism in government and authority that is already quite widely felt. Worse, despite good intentions, encouraging people to be "alert", rather than alarmed, may well further erode the very social bonds of ordinary human trust we need to depend upon if we are truly to be resilient as a society.

As identified earlier, the usual list of measures taken to enhance social resilience since September 11th 2001 consists amongst others of the need for better surveillance and intelligence, more effective models for predicting behaviour, new detection equipment and protective clothing, alternative modes for imparting information through "trusted" sources, as well as new structures of government and integrated response systems.

None of these serve to shore up ordinary social bonds and hence human and societal resilience. By encouraging the dominant paradigm of risk management in our understanding both of terrorism, as well as how to respond to it, we are encouraging a suspicion of others that effectively pushes people further apart and accentuates existing trends towards social atomisation. We have created a new bureaucracy but, as the figures show, we have failed to address the underlying insecurities [38].

Above all we have focused solely upon the form that terrorism now takes in the modern world – that relating in some increasingly tangential way, to Al Qa'ida – and largely ignored its content – a vehement anti-Americanism that rejects modernity and progress.

This reveals the real complacency of the dominant responses. One hardly needs to leave the West, to discover a whole host of other voices also expressing a hatred for America and progressive enlightenment values. This division is internal rather than external. Islamist terror is merely its most visible manifestation. Once "Stupid White Men" had become a best-seller on both sides of the Atlantic we should have been alert to a certain degree of cultural self-loathing at home [39].

Timothy McVeigh and the Aum Shinrikyo cult, pointed to our ability to create home-grown nihilist terrorism. It is well worth reminding ourselves that the 19 hijackers from September 11th 2001 had themselves all spent considerable time in the West, imbuing our values – or lack of them – and had largely been educated here.

Terrorism in every age reflects the dominant values of the most advanced societies. In the age when Western countries advanced and defended the sovereign rights of independent nation states, terrorists fought national liberation struggles. Today, in an age when it is not so clear what we truly believe in, we find terrorists that declare no aims and profess no responsibility for the carnage they create. Maybe it is time we examined ourselves more deeply rather than the final outcome of such values.

Cultural confusion as to who we are, what we are for and where we are going will undermine our attempts at instituting social resilience. Society today is less coherent than it was a generation or more ago, it is also less compliant, but above all it is less confident as to its aims and purposes. This will not be resolved by training ourselves to respond to disasters, but by a much broader level of debate and engagement in society, not just relating to terrorism and other crises, but to far broader social issues.

7. Social Solutions

Historical comparisons of disaster, such as responses to the Second World War "Blitz", or to past episodes of flooding and epidemic disease, reveal a number of important lessons for today. Not least is the extent and depth of social bonds and engagement at those times. During the war, there was a clear sense of the need to carry-on with normal life and everyday roles and responsibilities, rather than developing some kind of "shelter-mentality" [40], as is now encouraged through talk of stocking-up on batteries and fresh water.

However, the most striking change over the last fifty years has been in how we assume that ordinary human beings will react in a crisis. Beyond the grossly distorted belief in the likelihood of panic lies a more subtle, yet unspoken shift in cultural assumptions, that in itself undermines our capacity to be strong. That is, that in the past, the assumption was – on the whole born out by actual human behaviour – that people were resilient and would seek to cope in adverse circumstances.

Today, there is a widespread presumption of human vulnerability that influences both our discussion of disasters well before they have occurred, and that seeks to influence our responses to them long after. A new army of therapeutic counsellors and other assorted professionals are there to "help" people recover [41]. This presupposes our inability to do so unaided. Indeed, the belief that we can cope, and are robust, is often presented as outdated and misguided, or as an instance of being "in denial".

In some ways, this latter element, more than any other, best exemplifies and clarifies some of the existing confusions and struggles that lie ahead. If self-reliance is old fashioned and help-seeking actively promoted, for whatever well-intended reason, then we are unlikely to see a truly resilient society emerge.

This cultural shift is reflected in the figures that show that whereas in the United Kingdom, in the period of trade union militancy and unrest known as the "Winter of Discontent" of 1979, there were 29.5 million days lost through strikes, in 2002 there were 33 million days lost through stress [42].

We have shifted from being active agents of history to becoming passive subjects of it. This may suit social leaders lacking a clear agenda or direction. It may indeed be easier to manage the sick than those who struggle. Yet it also precludes the possibility of encouraging and establishing real resilience, resolve and purpose across society.

The standard way of dealing with disaster today is one that prioritises pushing the public out beyond the yellow-tape perimeter put up by the authorities [43]. At best the public are merely exhorted to display their support and to trust the professionals. Effectively, we deny people any role, responsibility or even insight into their own situation at such times. Yet, despite this, ordinary human beings are at their most social and rational in a crisis. It is this that should be supported, rather than subsumed or even subverted.

Handling social concerns as to the possibility of a terrorist attack is no easy feat. In part, this is because social fears today have little to do with the actuality, or even possibility, of the presumed threats that confront us. Rather, they are an expression of social isolation and mistrust, combined with an absence of direction and an elite crisis of confidence. Debates about the accessibility of technology and the reporting of science in the public domain have to be understood in this context, rather than being accepted and deliberated upon in their own terms.

The starting point to establishing real resilience and truly effective solutions will be to put the actual threat posed into an appropriate context. This means being honest as to the objective evidence, as well as being able to clarify the social basis of subjective fears. Engaging the public in a political debate over societal values may be a longer-term goal than dealing with any imminent terrorist threat, but it is necessary to inform our approach as a society.

The incessant debate as to the possibility and consequences of an attack using chemical, biological, radiological or nuclear weapons, is a case in point [17]. Whilst Western societies have debated such nightmare scenarios as if they were real, terrorists have continued to display their proficiency in, and proclivity to use, conventional weapons, such as high explosives, car bombs and surface-to-air missiles.

Above-all, if as a society, we are to ascribe an appropriate cultural meaning to the events of September 11th 2001 – one that does not enhance domestic concerns and encourage us to become ever-more dependent on a limited number of "expert" professionals who will tell the public how to lead their lives at such times – then we need to promote a far more significant political debate as to our aims and purposes as a society.

Changing our cultural outlook is certainly a daunting task. It requires people in positions of authority to clarify and agree on a common direction and then to win others to it. The reluctance to engage in this fundamentally political process and the clear preference to concentrate instead upon more limited, technical goals, leaves us profoundly ill-equipped for the future. It speaks volumes as to our existing state of resilience and may serve to make matters worse.

Bizarrely, few of the authorities concerned consider it to be their responsibility to lead in this matter. Nor do they believe such cultural change to be a realistic possibility. Yet, in the eventuality of a major civil emergency, they hope that the public will pay attention to the risk warnings they provide and alter their behaviour accordingly. By then it will be too late.

References

[1] Royal Society (UK). Making the UK Safer: Detecting and Decontaminating Chemical and Biological Agents. Policy Document 06/04. London: Royal Society; 2004.

[2] Kline M. Mathematics in Western Culture. Oxford: Oxford University Press; 1953.

[3] Boas Hall M. The Scientific Renaissance 1450-1630. New York: Dover Publications; 1994

[4] Pick D. Faces of Degeneration: A European Disorder, c.1848-c.1918. Cambridge: Cambridge University Press; 1993.

[5] Carr EH. What is History? New York: Vintage Books; 1967.

[6] Adorno TW, Horkheimer M. Dialectic of Enlightenment. New York: Continuum; 1976.

[7] Gillott J, Manjit K. Science and the Retreat from Reason. London: Merlin Press; 1995.

[8] Putnam R. Bowling Alone: The Collapse and Revival of American Community. New York: Simon & Schuster; 2000.

[9] Furedi F. Paranoid Parenting. London: Penguin; 2001.

[10] Furedi F. Culture of Fear: Risk-Taking and the Morality of Low Expectations. London: Continuum; 2002.

[11] Heartfield J. The 'Death of the Subject' Explained. Sheffield-Hallam University: Perpetuity Press; 2002.

[12] Giddens A. Runaway World: How Globalization is Reshaping Our Lives. London: Profile Books; 1999.

[13] Fukuyama F. The End of History and the Last Man. New York: Free Press; 1992.

[14] Huband M, Burns J, Krishna G. Chemical weapons factory discovered in a London flat. London: Financial Times; 8 January 2003.

[15] Noji E. Medical preparedness and response to terrorism with biological and chemical agents: present status in USA. International Journal of Disaster Medicine. 2003; 1:1: 51-55.

[16] HM Government (UK). Preparing for Emergencies. London: HMSO; 2004.

[17] Durodié B. Facing the possibility of bioterrorism. Current Opinion in Biotechnology 2004; 15:3: 264-268.
[18] Morse SS. Biological and chemical terrorism. Technology in Society 2003; 25:4: 557-563.
[19] Petro JB, Plasse TR, McNulty JA. Biotechnology: impact on biological warfare and biodefense. Biosecurity and Bioterrorism 2003; 1:1: 161-168.
[20] Beeching NJ, Dance DAB, Miller ARO, Spencer RC. Biological warfare and bioterrorism. British Medical Journal 2002; 324:7333: 336-339.
[21] Hyams KC, Murphy FM, Wessely S. Responding to chemical, biological or nuclear terrorism: the indirect and long-term health effects may present the greatest challenge. Journal of Health Politics, Policy and Law 2002; 27:2: 273-290.
[22] Hall MJ, Norwood AE, Ursano RJ, Fullerton CS. The psychological impacts of bioterrorism. Biosecurity and Bioterrorism 2003; 1:2: 139-144.
[23] Furedi F. Therapy Culture: Cultivating Vulnerability in an Uncertain Age. London: Routledge; 2004.
[24] Hassett AL, Leonard HS. Unforeseen consequences of terrorism: medically unexplained symptoms in a time of fear. Archives of Internal Medicine 2003; 162:16: 1809-1813.
[25] Wessely S, Deahl M. Psychological debriefing is a waste of time. British Journal of Psychiatry 2003; 183:1: 12-14.
[26] Lerner JS, Gonzalez RM, Small DA, Fischoff B. Effects of fear and anger on perceived risks of terrorism: a national field experiment. Psychological Science 2002; 14:2: 144-150.
[27] King NB. The influence of anxiety: September 11, bioterrorism, and American public health. Journal of the History of Medicine 2003; 58:4: 433-441.
[28] Malik K. Don't panic: it's safer than you think. New Statesman 2001; 14:67: 18-19.
[29] Durodié B. Cultural precursors and psychological consequences of contemporary Western responses to acts of terror. In: Wessely S, Krasnov V, eds. Psychological Aspects of the New Terrorism: A NATO Russia Dialogue. Amsterdam: IOS Press; 2005 in press.
[30] Durodié B, Wessely S. Resilience or panic? The public and terrorist attack. The Lancet 2002; 360:9349: 1901-1902.
[31] Furedi F, Roberts S. Disaster and Contemporary Consciousness: The Changing Cultural Frame for the Experience of Adversity. Draft Report available from the author; 2004.
[32] Durodié B. Is real resilience attainable? RUSI/Jane's Homeland Security & Resilience Monitor 2003; 2:6: 15-19.
[33] Cabinet Office (UK). Draft Civil Contingencies Bill, Consultation Document. London: HMSO; 2003.
[34] Furedi F. A sociology of health panics. In: Mooney L, Bate R, eds. Environmental Health: Third World Problems – First World Preoccupations. London: Butterworth-Heinemann, 1999.
[35] Cowan R. Attack on London is 'inevitable'. Manchester: The Guardian; 17 March 2004.
[36] Manningham-Buller E. The oversight of intelligence and security. Speech to the Royal United Services Institute, Whitehall, London; 17 June 2003.
[37] Blair T. Speech at the Lord Mayor's Banquet, Whitehall, London, 11 November 2002.
[38] Durodié B. Panic in the streets? New Humanist 2004; 119:3: 18-19.
[39] Moore M. Stupid White Men … and Other Sorry Excuses for the State of the Nation. New York: Harper Collins; 2001.
[40] Jones E, Woolven R, Durodié B, Wessely S. Civilian morale during the Second World War: responses to air raids re-examined. Social History of Medicine 2004; 17:3: 463-479.
[41] Furedi F. Therapy Culture: Cultivating Vulnerability in an Anxious Age. London: Routledge; 2003.
[42] Marsden C, Hyland J. Britain: 20 years since the year-long miners' strike. World Socialist Web Site, 5 March 2004. Available at: http://www.wsws.org/articles/2004/mar2004/mine-m05.shtml
[43] Glass T, Schoch-Spana M. Bioterrorism and the people: how to vaccinate a city against panic. Clinical Infectious Diseases 2002; 34:2: 217-223.

Science and Technology Policies for the Anti-Terrorism Era
Andrew D. James (Ed.)
IOS Press, 2006

R&D and the War on Terrorism: Generalising the Israeli Experience

Isaac BEN-ISRAEL, Oren SETTER and Asher TISHLER
Tel Aviv University, P.O. Box 39040, Ramat-Aviv, Tel Aviv, Israel

Abstract. Terrorism is by no means a new phenomenon. In recent decades, however, it has become a major security threat that requires a shift in the focus of the defence efforts of many developed countries to designing effective anti-terrorism strategies, and corresponding R&D strategies. This Chapter analyses the main characteristics of terrorist organisations with a view to identifying their main weaknesses. We find that the key weaknesses of terrorist organisations are their small size, their high vulnerability to any action against their leadership, and their high susceptibility to the use of high-tech intelligence gathering and processing systems. The Chapter recommends that the main anti-terrorism strategies should be active (pre-emptive) and focus on applying continuous pressure on key activists (including leaders) of terrorist organisations. Since such pressure is attainable only through the coordinated effort of a broad range of defence-related agencies, the emphasis of the anti-terrorism R&D strategy should be on investment in integrative technologies and in particular on long-term intelligence (LTI) systems.

1. Introduction

Terrorism is by no means a new phenomenon. In recent decades, however, it has become a major security threat that requires a corresponding shift in the focus of defence-related R&D efforts of many developed countries. Of the several economic models that deal with defence R&D (see, for example, [1, 2, 3, 4, 5, 6, 7, 8]), we focus on those dealing with the technologies and R&D that are required to effectively combat terrorist organisations.

Conceptually, we consider the war on terrorism as a game between a government, or several governments, and terrorist organisations. First, we discuss several characteristics of terrorist organisations and identify their major weaknesses and strengths. Next, we focus on the weaknesses of some of these organisations and identify technologies and actions that may be helpful in limiting their successes or, better, drastically reducing their capabilities and relevance. The appropriate R&D programmes to support these technologies and actions are then derived, and possible means of enhancing the efforts of civilian industry to contribute to this R&D effort are discussed. Recognising that we are in a game setup, we suggest examining how the relevant R&D and actions designed to be taken by governments can favourably (or unfavourably) change the equilibrium of the game and even the game itself [9, 10]. Finally, we discuss the two types of anti-terrorism actions and policies that can be adopted: (i) active actions and policies aimed at pre-emptive attacks on these organisations and, (ii) passive policies and actions aimed at defending the population from terror attacks. Clearly, these actions and policies can be complementary and may overlap. In this Chapter we consider both types of actions and policies, but emphasise the active (pre-empting) approach to the war on terror.

This Chapter draws on the Israeli experience. In the four years after the beginning of the second (Suicide Terrorism) *Intifada* in 2000, Israel faced unprecedented levels of terrorism, especially in the form of suicide bombers. In 2004, after a four-year concerted effort, Israel succeeded in bringing the level of terrorism down to its pre-2000 level. The Chapter discusses the main lessons that can be learned and applied from this experience in the context of R&D. These may be summarised as follows. First, the top priority of counter-terrorism measures should include actions (including R&D efforts) directed against the higher echelons of terrorist organisations. Second, viewing acts of terrorism as the end result of a production line, the priority of counter-terrorism against a certain link in this line should be proportional to its place in the line: the closer it is to the origin, the more effective will be its neutralisation. Third, it is more effective to prevent terrorists entering a country than trying to stop them when they are already inside and undertaking their deadly operations. Detection technologies on the border, creating a virtual fence, are more cost-effective than detection technologies against terrorists in the streets or shopping malls.

2. The Roots and Characteristics of Global Terrorism

In our view, current Islamic terrorism is *not* a war between religions. Osama Bin-Laden does not want to convert the Americans nor does President Bush want to convert Muslims (though the term "crusaders" has been heard). Islamic terrorism does not seem to have a material cause either and it is not being used as a way to improve the economic situation of the Muslims. Though the issue of oil supply certainly played a role in the Bush administration's decision to attack Iraq, it was not the reason for its launching the war on terror, nor was it the reason for Bin-Laden's attack on the World Trade Center and the Pentagon. Samuel Huntington's concept of a clash of civilizations might be more helpful here [11]. However, it is still insufficient. What is a "civilization"? Why is terrorism not supported by the Indian or Chinese civilizations?

What drives terrorism? We think that a possible answer is that it is an internal clash between supporters of open societies and opponents of open societies. This kind of clash exists in many societies from Afghanistan to the United States, Italy or Israel. On one side there are people who view modernisation, change, innovation, technology and so forth as good and desirable goals. For them these terms express positive attributes. On the other side there are those who view tradition as the most important value of society. In their eyes, technology is evil: TV corrupts the souls of the young and cellular phones pose a threat to the old system. They fear that the end result of modernisation will be the destruction of many of the old values. For them words like change and modernisation carry an evil message and spell the decline and fragmentation of the family, and revolt against authority in general. The United States is seen as the "the big devil" because it spreads this culture of modernisation.[1] Bin-Laden acted against the USA because he understood that the only way to save his Islamic world, and immunise it against change and modernisation, was to quarantine the progenitor of this new culture outside the physical space of the Muslim world.

At the end of the day one has to choose: does one desire to live in an open society (where alternatives are rationally debated and decided) or in a closed society (where keeping the old order is the ultimate maxim)?[2] Very few societies have chosen the first way. Most of the world still belongs to the second group. It should be emphasised here that the measure of openness in a particular society is a matter of degree. It is not a black and white distinction. There are societies that are more open than others. Even within the Islamic world there are societies that are very closed (like Saudi Arabia and Syria) and those that are relatively open (like Indonesia and Turkey). Democracy, according to this view, is only one aspect of an open society, and it too is a matter of degree. North Korea, Iran and Syria belong to the same group because they are closed societies (and not because they represent the same "civilization").

We have to distinguish clearly between the goal and the means in this context. The goal is to keep the Islamic world closed. The means is terrorism. It should be emphasised that this goal is not necessarily a “Muslim” goal and we do not think it is shared by the majority of the Muslim world. It is the leading ideology of a small group, and it could happen anywhere in any society and under any religion. The reason why Bin-Laden chose terrorism as a tool to achieve his goal is that open societies are vulnerable to terror. Open societies allow free movement of individuals and goods, they respect human rights and freedom of speech and expression. Terrorism is only a tool in a war to keep the old values unchanged forever.

2.1 Global versus Local Terrorism

Some of the centres of gravity of terrorist action around the globe are closely connected to "local" conflicts, such as the conflict between the Palestinians and Israel, the Chechens and Russia and Pakistan and India in Kashmir. Is there a difference between local and global terrorism?

Some differences are evident. While local terrorism is usually geographically confined, global terrorism knows no borders. Local terrorism has clear immediate political demands which can, in principle, be settled politically by negotiations between the two sides. Global terrorism is motivated by an ideological vision which has no clear-cut demands for immediate results.

Nevertheless, the two types can interact and support each other. This is clearly the case when global terrorism offers help to local organisations. Sometimes the two work together locally, as is the case in Israel, where more than 90 percent of the suicide attacks in the last four years were carried out by Hamas and Islamic Jihad. Their goal is not an independent Palestinian state in the occupied territories. They act on a pan-Islamic platform intent on restoring the rule of the Koran in the whole space that historically belongs, according to their view, to the Muslims.[3] Indeed, every time the Israeli government and the Palestinian Authority (PLA) have come close to some kind of an agreement, Hamas and Islamic Jihad have increased their acts of terrorism.

The goal of having an independent Palestinian state is no closer today than it was before the start of the 2000 *Intifada*. However, Hamas thinks it is a success because its goal is different: before the *Intifada* only 17 percent of the Palestinians supported Hamas, and the number today is around 40 percent.[4] This "forced" the PLO to join the suicide attacks (in order to retain the support of the people) and the rest is history: the process of bridging the differences between Israel and the Palestinians has been practically dismantled.

Governments in open societies cannot give in to terrorism and terrorism has to be reduced significantly before real negotiations can start on solving the local "political" conflict. This is true for global terrorism as well. Global terrorism is not a threat that can be completely eliminated. The “war” against terrorism is not a classic war (like the war between two rival armed forces). It is more like the war against crime. Terrorism, like crime, will not disappear totally. Therefore, the goal of the “war” against terrorism should not be to eliminate it but to reduce it to a bearable level.

However, there is much that can be done at the global, multi-lateral level. Terrorism has global roots, and these can be greatly reduced or cut off through cooperative efforts involving a sufficiently large number of countries.

2.2 Characteristics of Terrorist Organisations

Success in the invisible and borderless war against terrorist organisations requires, among other things, the identification of the main characteristics of these nebulous entities. Terrorist organisations do not have well-defined structures or sizes, permanent bases, or long-term detailed plans and updated lists of relevant actions and targets. They operate on the basis of a broad definition of possible terror activities and general targets, and usually possess minimal intelligence on the targets against which they decide to take action.[5] Most of this intelligence is derived from publicly available data and common knowledge rather than the meticulous and detailed data gathering process that is common to modern intelligence organisations. Terrorist organisations are knowledgeable about the various weak points of the regimes that they are fighting against. They rarely have detailed operational plans at any given point of time, tending to execute a terror act when they are ready, and not necessarily when their targets are weak or particularly vulnerable.

Terrorist organisations do not have unlimited resources. To execute most of their operations they need support from ideological groups and sympathy from certain sections of the civilian population. They are likely to have some training with small arms and possibly some of the more sophisticated weapon systems that are available in the market. Rarely, if ever, are they capable of executing large-scale operations by large, coordinated units, using modern weapon systems, though they may posses some logistical and information systems. Large-scale terror operations require a wide base of support in logistics, training, planning, finance and communications, which can only be acquired through substantial external support. For this support to be effective it must be located in countries that are friendly to the terrorist organisation or at least not hostile, and it is more effective when the host government actively supports the terrorist organisations or at least shields them from outside pressures. Clearly, there may be symbiotic relationships between the terrorist organisation and the host government – possibly through some of its institutions, or some local power group fighting for control of the country or some of its resources. In return for their hospitality, the terrorist organisations may supply their hosts with expertise, weapons, logistics, know-how and other inputs that are important to them.

Finally, successful terrorist organisations rely on intelligent, ideological, possibly intellectual, and well-functioning leadership. The importance of this leadership is apparent when one looks at the execution of a terror act as the outcome of a "production line". Focusing on those carrying out the acts of terror (the suicide bombers, for example) is wrong. One has to bear in mind that these suicide bombers are the end result of a long chain of activity. One has to decide to carry out the act of terrorism. It has to be planned, but not before appropriate intelligence is collected about the target, its protection and a route to the target that is not too vulnerable to early detection. The explosive device has to be prepared. A volunteer has to be recruited and instructed. Transportation has to be arranged for the explosive device to reach the volunteer and for both to reach the selected target. All these actions need to be taken care of and coordinated by an organisation. The crucial point here is that the key people in a terrorist organisation are very few. These are very small organisations. The number of key activists in Hamas, for example, who are actually engaged in preparing an act of terror, is only a few hundred (not counting the passive supporters). One only needs to neutralise 20-30 percent of them for the organisation's "production" of acts of terror to drop significantly.

The next section details the Israeli experience in combating terrorism, based on the logic described above. This experience is then generalised in later sections, with the aim of devising a global anti-terrorism policy.

3. The Israeli Experience

Israel has been a terrorist target since it became a state in 1948. Starting from the 29th September 2000, however, the rate of terror activity climbed to an unprecedented level. Figure 1 describes the number of fatal casualties (per quarter) between 2000 and 2004.

Although suicide terrorism has been used in the past, it was during these four years that it emerged as the main tool employed by the terrorists. Between September 2000 and September 2004, some 553 suicide attacks were attempted; 135 of them achieved their goal, killing 880 non-combatant innocent civilians, including women, children, and elderly people. By late 2004 and after four years of combating terrorism in Israel, it could be said with caution that clear signs of success were appearing: the rate of terrorism had dropped back to pre-*Intifada* figures, as can be seen in Figure 1. Figure 2 provides some of the more important statistics relating to this phenomenon.

Generally speaking, the turning point occurred after the war on terror was "separated" from the issue of a political agreement with the Palestinian leadership. In April 2002, after Israel suffered 140 fatal causalities in the previous month due to suicide terrorism, the wish to reach an agreement with the Palestinian Authority (PLA) was put aside (Arafat was declared to be "non-relevant"), and Israel launched a full-scale campaign on the terrorist organisations, concentrating on Hamas and Islamic Jihad. This campaign started by expanding the intelligence coverage, mainly by entering the occupied areas that had come under the control of the PLA after 1993. This, together with the security "fence" resulted, a year later, in an increase in the percentage of aborted terrorist acts from 40 percent (before April 2002) to over 80 percent (see Figure 3). Though the decrease in the number of successful terrorist attacks is attributable mainly to the increased percentage of aborted attacks, the decrease in the number of terrorist attempts from mid-2003 is a direct result of pre-emptive strikes, namely, the so-called "surgical elimination" of numerous key members of the two main terrorist organisations.

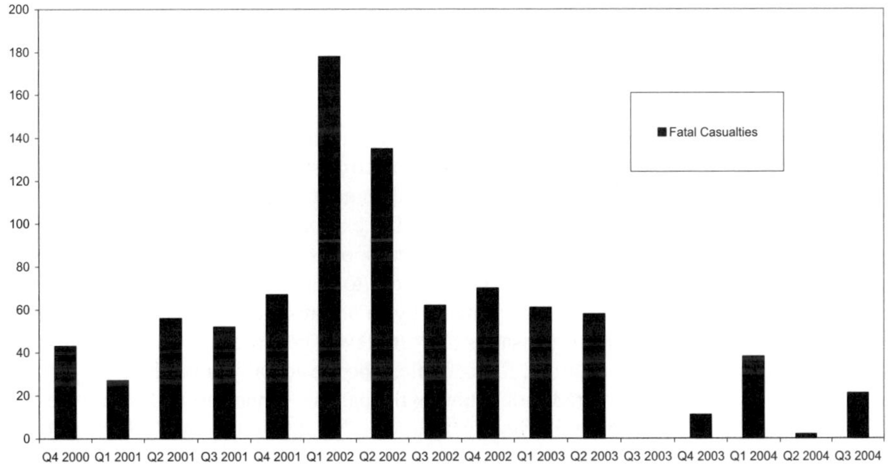

Figure 1: Distribution of Fatal Terrorism Causalities 2000-2004

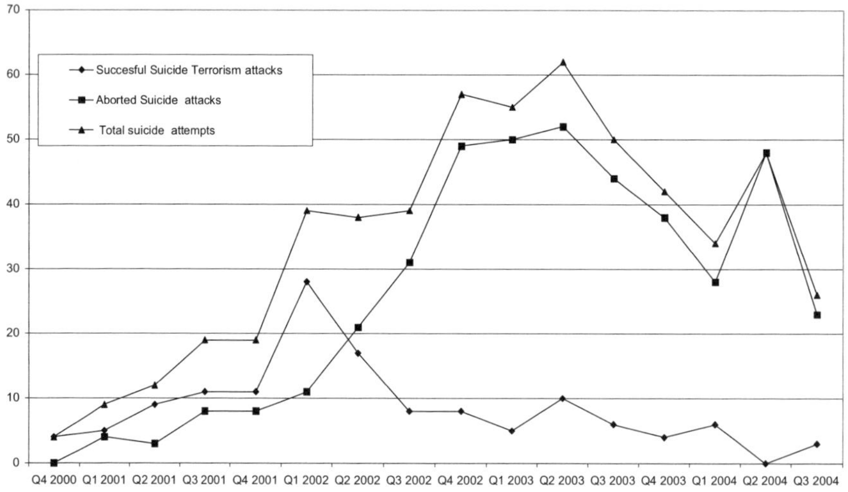

Figure 2: Suicide Terror Attacks

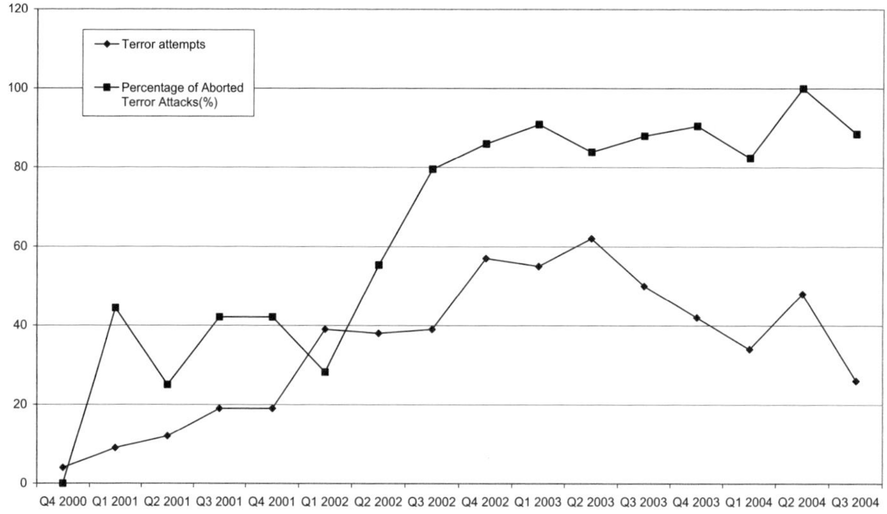

Figure 3: Percentage of Aborted Terror Attacks

These terrorist groups are rather small. They enjoy the support of the street but this support is usually passive. The number of people actively involved in terrorism in any such organisation is no more than a few hundred including the planners, the engineers and technicians who produce the explosive belts, those who arrange the transportation, the intelligence gathering activists and the decision-making leadership. It is a small enough number to allow the elimination of its major figures. When the rate of elimination of key people in a terrorist organisation reaches 20-30 percent, it greatly reduces its function as a producer of terrorist acts. Increasingly, the organisation starts occupying itself with a different kind of war: a war for its own survival.[6]

Looking at Figure 1, one may ask how Israel had succeeded in reducing the level of terrorism. Does Israel have the technology to detect explosive belts being carried by suicide bombers in the streets? Is the technology used only in vital locations such as air and maritime ports and check points or more widely at the targets favoured by suicide bombers, such as restaurants, buses and shopping malls? What is the range of detection? Does it also include means to neutralise or activate the explosive belts? Whatever the desirability of such technologies, the answer to these questions is generally negative. Though there are some technologies that perhaps have the capabilities described above, they are not mature enough and need a long R&D period before they can be fielded.

So, how did Israel succeed in reducing the level of terrorism so substantially? Was it done without technology? The answer is again negative. It would not be an exaggeration to say that science and technology and, particularly, R&D played a major role in suppressing terrorism in Israel. This is especially true in the case of suicide terrorism. Science and technology has been directed not towards the end points (suicide bombers) but against the higher echelons of the terrorist organisations.

4. Anti-Terrorism Policies and Actions

Based on the Israeli experience, we now outline a general policy to combat terrorism, and particularly the required R&D policies. At the outset, it should be noted that it is very difficult for governments to use passive defence policies to completely and effectively prevent all terror activities everywhere in the country. There are several reasons. First, terrorist groups are not particularly concerned with the specific members of the population that they hit. Typically, anyone will do. Second, the fuzzy nature of terrorist organisations, which serves, among other purposes, to avoid detection by the government, prevents them from having long-term detailed plans for terror activities. Rather, the decision to hit a specific target is often made on relatively short notice, making it difficult for governments to predict the time, location and method of forthcoming terror activities. The following lesson from the Israeli experience is thus of central importance: when a terrorist (e.g., a suicide bomber) is already in town, it is too late. There are some measures that can be taken to reduce the number of casualties, but there is, practically, no way to prevent him from blowing himself up.

Hence, an active approach has to be taken of fighting terrorism at its source. Clearly, in addition to the obvious and necessary political, social and economic means, effective ways to fight terror are to exploit the terrorist organisations' weaknesses. First, we have already noted that successful terrorist organisations are heavily dependent upon their leadership and as in other organisations such as firms or governments, that leadership is very difficult to replace (see, among others, [14, 15, 16]). Second, terrorist organisations are networked, and thus rely heavily on lines of communication that are vulnerable to eavesdropping, monitoring, and even deception. The terrorist organisations, for their part, have few, if any, means to effectively prevent or even know about such scrutiny. Hence, an effective anti-terrorism strategy applies continuous pressure on the leadership of terrorist organisations by using long-term information gathering, high-tech surveillance, powerful data-analysis technologies, and intelligent system analyses of their activities and structures. The implementation of such a strategy to supply sufficient information for continuous and effective action against existing and potential terrorist organisations is time consuming and requires substantial resources.

4.1 The Role of Long-Term Intelligence

We define long-term intelligence (LTI) as the continuous gathering, quantification and analysis of information obtained over a long period of time, in different spaces (publicly available, electronic and so forth) and locations. This will use various means including human intelligence (HUMINT) and signals intelligence (SIGINT). The need for LTI stems from the need to know and understand the leaders of terrorist organisations as well as potential future leaders. Using a long-term approach allows attention to be focused on the more important leaders, and not just on urgent cases. Effective information gathering requires the collection, organisation and storage of data according to measures that can be continuously analysed and compared (not necessarily in real time) across individuals, time, situations, and locations.

The active (pre-emptive) approach to fighting terrorism focuses the intelligence on people (leaders, commanders and main operators), while the passive approach shifts the intelligence focus to locations and activities. It would therefore seem to be much simpler to collect and maintain the relevant information on the relatively small set of leaders, commanders and main operators, than on the set of all possible terror activities at all possible locations (see [7], for a formal model describing this asymmetry). In addition, the active approach shifts most of the operations from the country's own civilian population and physical space to those controlled by or sympathetic to the terrorist organisations. Thus, active policies and actions should get the lion's share of attention and budgets in any anti-terrorism activities.

In summary, a key concept in the war on terrorism is the use of cutting-edge technologies. While the apparatus of terrorism is usually low-tech, it would be a huge mistake to think that the answer to terrorism should also be low-tech, and to subscribe to the notion that technology is good only for classic wars between armies. The Israeli experience proves the opposite: the more technological the war against terrorism becomes, the better are the results. The reason for this is, again, that terrorist organisations are small. Unlike sovereign states, they do not have enough men and resources to cope with high-technology responses.

4.2 The Organisation and Operation of the Anti-Terror Apparatus

We have argued that the war on terrorism cannot be local; it must be global and comprehensive. It is important to note that a large percentage of the means and information to fight terror already exist in various organisations such as the military, police and other security-related organisations [17, 18, 19, 20]). However, none of these organisations possesses all of the means or capabilities to fight terrorism. It seems clear, for example, that the war on terror cannot be managed solely by the military. The military's focus and responsibility are on all-out wars, making it incapable of providing the necessary managerial attention and, hence, financial and human resources to the war on terrorism [21, 22]. Moreover, the organisational culture of the military is set to deal with and solve large-scale problems, not a series of very heterogeneous and ever-changing situations.

It follows that a new body entrusted with the responsibility for leading the war on terror has to coordinate, or control, a variety of organisations that have the responsibility for other tasks (some of which are correlated to the prevention of acts of terror) and will, most likely, object to any intrusion into their natural environment and inroads into their scarce financial and human resources. Similar to the Department of Homeland Security in the USA, this body should therefore have its own resources as well as the formal obligation and authority to conduct policy, control resources belonging to other bodies, develop its own technologies and use them as needed. In brief, our discussion and the analysis contained in a number of studies [20, 23, 24, 25] suggest the following:

1. The war on terrorism should employ a diverse set of actions and technologies and be carried out by several existing and possibly new organisations at various locations. There is no single short-term winning policy or action that can effectively defeat terrorist organisations.
2. Centralised long-term intelligence (LTI) should be at the heart of anti-terrorism activities. That is, LTI should integrate the multitude of activities and policies to effectively fight terrorist organisations.

These two points lead us to suggest a methodology based on the concept of integrative technologies to describe and evaluate an effective way to fight terrorist organisations.

5. R&D and the War on Terrorism

5.1 Integrative Technologies – Network Effects in the War on Terrorism

Setter and Tishler [8] define integrative technologies as "information and communication technologies that enable separate individual systems to work in a joint, coordinated, and synergistic fashion as a single holistic system". Typical examples of integrative technologies in the military are Command and Control (C^2) systems, global communications networks, cross-service information systems, and integrative intelligence systems. The value of these technologies stems not only from the network they create, which can be obtained using other simple means (tactical radios and telephones, for example) but also from the quality of integration they provide. The higher the quality of integration, the more effective is the overall operation of the networked units.

Integrative technologies exhibit three key characteristics. First, they have a defence-wide effect because increased quality of integration benefits the entire defence apparatus, and not just any particular component of it. Second, they have the capability of exhibiting an exponential network effect because they influence the quality of the interactions among systems, and the number of connections is exponential in the number of connected systems (size of the network). Finally, integrative technologies are a relatively new phenomenon, and their development is likely to exhibit increasing returns.

Many technologies have evolved recently that improve the quality of defence integration, and change its characteristics. The U.S. military, a global technological leader, accords very high priority to integrative technologies and is developing concepts that place such technologies at their core, not least through "Network Centric Warfare" [26, 27]. The 2001 Quadrennial Defence Review Report concluded that future U.S. defence funding will focus on achieving integrated joint "end-to-end Command, Control, Communication, Computers, Intelligence, Surveillance, and Reconnaissance capabilities" [28]. The application of these concepts to the war on terror is straightforward, and should be based on a broad range of programmes and technologies.

Setter [29] and Setter and Tishler [8] develop an analytical framework that captures the defining characteristics of integrative technologies for the military, and provides a simple budget allocation procedure that yields a unique optimal solution. The main results of this model are as follows. First, the optimal investment in integrative technologies grows with the number of systems. Second, depending on a country's defence budget level, its optimal investment in integrative technologies is either zero or large. There is no benefit from "too small" investments in this type of technology, though when the investment is large, the return (in operational terms) is very high.

The framework developed by Setter and Tishler can be easily applied to R&D decisions in the context of the war against terror. While conventional wars typically require the integration of three services (army, air force and navy), the war on terror requires integrating a multitude of defence-related agencies and organisations (military services, special forces, police forces, intelligence agencies, R&D agencies, defence contractors, and so forth). Hence, if the US spends approximately 20-25% of its defence R&D budget on integrative technologies (see [29] for details), then the relative investment in integrative technologies for fighting terror should be even greater. Hence, proper integration of the various relevant bodies is the key to an efficient and effective operation to vastly reduce, and possibly eliminate, the threat of terrorist organisations.

Long-term intelligence is clearly at the heart of a successful integration because it is required for developing the capability to collect, merge and fuse data obtained from various sensors (sometimes, but not always, in real time) to develop information relating to a particular person, group, mission, task or target.

5.2 R&D and Long-Term Intelligence

Accordingly, R&D activities should support the LTI approach. Some of the sensors and short-term intelligence are already available, but the R&D programmes that support them are fairly unique to the defence forces [19, 20]. Additional R&D programmes that are needed to support integration and the LTI approach are as follows:

1. Automatic, computer-based, management information systems (MIS) to quantify and analyse data (sequentially or sporadically) over long periods of time.
2. Data collection from civilian as well as defence sources.
3. Methodologies to quantify, fuse, and store the relevant data.
4. Data collection in transient locations.
5. Computer-based methodologies to understand the various leaders and commanders of the terrorist organisations ("customers"). These methodologies should be similar to those used in marketing, which are aimed at dealing with a large number of heterogeneous customers with varying requirements.
6. Data mining methods and methodologies (such as discrete choice models and multivariate analysis models) to estimate probabilities of different behaviours.
7. Models to assess (by computers) the situations at the relevant locations (spaces).
8. Automatic flags to indicate alarming behaviour.
9. Methodologies to evaluate and assess fuzzy situations, and automatically compare situations, individual and group behaviour over time.
10. Proper and independent controls on the organisations and individuals in charge of the surveillance, quantification and analysis activities.[7]
11. Methodologies to ensure that the right data are properly collected and are not improperly used, to prevent LTI from intruding on various notions of civil rights.[8]
12. Technologies to identify and detect terrorists and potential terrorist activities (at borders, in population centres, airports, communication systems and so forth) should operate from a distance and be unobtrusive, passive, silent and unrecognisable in order to avoid major interference with the civil rights of the country's residents, prevent detection by potential terrorists, and minimise recognition by anyone who is not a member of the security forces.

Unlike past defence technologies that have been developed almost solely by defence contractors many of the technologies required for the war on terror may be developed and are indeed being developed by civilian R&D companies. For example, data mining tools are being developed for marketing purposes, aiding firms in understanding the characteristics of their customers based on their past shopping behaviour. Moreover, data fusion and storage tools are being developed for many civilian applications and the concept of LTI could be used for business intelligence, allowing predictions of top-management decisions of competing firms. Finally, long-range passive sensors are being developed for a variety of civilian security applications.

6. Summary

The war on terrorism is perceived by many as the major security challenge to Western countries for the foreseeable future. Crafting effective anti-terrorism strategies, and corresponding R&D strategies, is therefore of clear importance. In this Chapter, we have analysed the main characteristics of terrorist organisations in order to focus on their main weaknesses. We find that the key vulnerability of terrorist organisations is their leadership, and that they are also susceptible to the use of high-tech intelligence gathering and processing systems.

The Chapter thus recommends that the main anti-terrorism strategies should be active (pre-emptive) and focus on applying continuous pressure on the leaders and higher echelons of terrorist organisations. Since such pressure is attainable only through a coordinated effort by a broad range of defence-related agencies, the emphasis of any anti-terrorism R&D strategy should be to invest in integrative technologies and in particular in LTI (long-term intelligence) systems.

Summarising the main results of the Israeli experience, it is clear that it is more effective to act against the leadership of terrorist organisations than against the activists in the field, and it is more effective to act against key activists playing a role in producing terrorism than against the terrorists who actually carry it out. By the same principle, it is more effective to prevent a terrorist from entering one's population centres than to attempt to stop him while he is already carrying out his "mission". Detection technologies on the border, creating a virtual fence, are more cost-effective than detection technologies against terrorists in the streets or shopping malls.

Thus, in a nutshell, three fields emerge as relevant for the science and technology community: intelligence (internal and external); surgical capabilities (concentrating on terrorist leadership and key activists); and perimeter security (fence or wall).

Notes

1. In the Middle East, Israel is usually referred to as the "little devil", because it is perceived as the local agent of American culture in the region.
2. This definition is in the spirit of Karl Popper, who deliberated on these issues in the 1940s. See [12] and [13].
3. In the words of Hamas: "The Islamic Resistance Movement believes that the land of Palestine is an Islamic Waqf consecrated for future Muslim generations until Judgment Day. It, or any part of it, should not be squandered: it, or any part of it, should not be given up. [...]This is the status [of the land] in Islamic Sharia (law), and the same goes for all lands conquered by Muslims by force, during the times of (Islamic) conquests, and made thereby Waqf lands upon their conquest, for all generations of Muslims until the Day of Resurrection". Source: http://www.hamasonline.com

4 The data is taken from the polls conducted by Khalil Shikaki and published by the Palestinian Center for Policy and Survey Research (PSR). http://www.pcpsr.org/survey/index.html

5 Experienced military officers will describe most of these actions as fairly easy for well-trained soldiers to execute. Thus, the intelligence available to the terrorist organisations is sufficient for their purposes.

6 Elimination can be accomplished either by arresting the terrorists (as is usually done in the West Bank, which has been reoccupied by the Israeli Defence Force), or by killing them when arrests are not possible (as in Gaza).

7 For example, it may be useful to consider the futures markets that exist for products and services as different as physical commodities (e.g. grain or lumber) and financial indices (e.g. foreign currencies or interest rates). Such markets, which have been developed in recent years to predict various future events (including the prediction of terror activities and the likelihood of influenza), are accurate for four reasons: (1) they aggregate information from all participants, each of whom has different information about the issue in question; (2) they provide incentives to encourage knowledgeable participants to reveal true information in their trades; (3) they provide feedback to participants – through market prices, traders learn about the beliefs of others and are motivated to collect more information; (4) all trades are anonymous – thus, they can signal information through the market that they might not want to announce publicly, or in person, to their peers (military officers and analysts, for example). The Iowa Electronic Market (IEM) is one example of a successful prediction market [30].

8 Treating terror acts and terrorists by using the conventional legal system is ineffective and possibly unfair to the rest of society. Thus, the legal system should also be prepared to deal with terrorists and, particularly, their leadership, effectively and fairly.

References

[1] Stoneman P. The Economic Analysis of Technology Policy. Oxford: Clarendon Press; 1987.

[2] Rogerson WP. Quality vs. quantity in military procurement. American Economic Review 1990; 80: 83-92.

[3] Hirao Y. Quality versus quantity in arms races. Southern Economic Journal 1994; 2: 96-103.

[4] Sandler T, Hartley K. The Economics of Defense. Cambridge: Cambridge University Press; 1995.

[5] Lichtenberg FR. Economics of defence R&D. In: Hartley K, Sandler K, editors. Handbook of Defence Economics. Vol. I. Amsterdam: Elsevier Science; 1995.

[6] Garcia-Alonso MC. Price competition in a model of arms trade. Defence and Peace Economics 1999; 10: 273-303.

[7] Trajtenberg M. Defense R&D policy in the anti-terrorist era. Working Paper no. 9725. Cambridge (MA): National Bureau of Economic Research; 2003.

[8] Setter O, Tishler A. The role of integrative technologies as a "force exponent" on military capability. Presented at The Eighth Annual Conference on Economics and Security; 2004 June 26-28. University of the West of England. Bristol.

[9] Kagan K, Tishler A, Weiss A. On the use of terror weapons vs. modern weapon systems in an arms race between developed and less developed countries. Presented at The Second International Conference on Defence, Security and Economic Development; 18 - 20 June 2004. TEI of Larissa. Greece.

[10] Kirkpatrick DLI. Trends in the costs of weapon systems and the consequences. Defence and Peace Economics 2004; 15: 259-273.

[11] Huntington S. The Clash of Civilizations and the Remaking of World Order. New York: Simon & Schuster; 1997.

[12] Popper K. The Open Society and its Enemies. London: Routledge & Kegan Paul; 1945.

[13] Popper K. The Poverty of Historicism. Second Edition, London: Routledge & Kegan Paul; 1961.

[14] Castanias RP, Helfat CE. Managerial resources and rents. Journal of Management 1991; 17(1): 155-171.

[15] Finkelstein S, Hambrick D. Strategic Leadership: Top Executives and their Effects on Organizations. St. Paul, Minneapolis: West Publishing Company; 1996.

[16] Carmeli A, Tishler A. The relationships between intangible organizational elements and organizational performance. Strategic Management Journal 2004; 25: 1257-1278.

[17] Barzilay A. The digital revolution of the IDF. Haaretz; 8 August 2003 (in Hebrew).

[18] Ben Israel I. The revolution in military affairs and the operation in Iraq. In: Feldman S, editor. After the War in Iraq: Defining the New Strategic Order. Brighton: Academic Press; 2003.

[19] James AD. U.S. Defence R&D Spending: An Analysis of the Impacts. Rapporteur's report for the European Union Research Advisory Board. Manchester: PREST, University of Manchester; 2004.
[20] Mendelevitch H. The absence of mature technology for fighting suicide bombers as a symptom of organizational failure. Mimeo; 2004 (in Hebrew).
[21] Christensen CM. The Innovator's Dilemma. Second Edition. New York: HarperBusiness; 2000.
[22] Alic JA, Branscomb LM, Brooks H, Carter AB, Epstein GL. Beyond Spinoff: Military and Commercial Technologies in a Changing World. Boston: Harvard Business School Press; 1992.
[23] American Association for the Advancement of Science [AAAS]. Making the nation safer: the role of science and technology in countering terrorism. In: AAAS Science and Technology Policy Yearbook 2003. Washington DC: AAAS; 2003. Available at: www.asss.org/spp/yearbook/2003/ch23.pdf.
[24] Kam E. Report of the Committee to Assess the Intelligence System. Strategic Assessment. Tel Aviv University: Jaffee Centre for Strategic Studies; 2004.
[25] Eilam U. Technology to fight terror. Strategic Assessment. Tel Aviv University: Jaffee Centre for Strategic Studies; 2004.
[26] Etter DM. Defense science and technology. In: Teich AH, Nelson SD, Lita SJ, editors. AAAS Science and Technology Policy Yearbook 2002, Washington DC: American Association for the Advancement of Science; 2002.
[27] Alberts DS, Garstka JJ, Stein FP. Network Centric Warfare. Washington DC: Department of Defense Command and Control Research Program; 1999.
[28] Department of Defense [DOD]. Quadrennial Defense Review Report. Washington DC: DOD; 2001.
[29] Setter O. Defence R&D in the Information Age: Analysis of Budget Allocation Decisions. Unpublished doctoral dissertation. Tel Aviv University; 2004.
[30] Berg J, Forsythe R, Nelson F, Rietz T. Results from a dozen years of electronic futures markets research. In Plott C, Smith V, editors. The Handbook of Experimental Economics Results. New York: Elsevier Press; 2005.

Science and Technology Policies for the Anti-Terrorism Era
Andrew D. James (Ed.)
IOS Press, 2006

Making the UK Safer: Detecting and Decontaminating Chemical and Biological Agents

Alastair HAY
Molecular Epidemiology Unit, School of Medicine,
University of Leeds, LS2 9JT, United Kingdom

Abstract. Most attacks by terrorists have involved the use of conventional munitions including explosives. It is likely that explosives will remain the weapon of choice by terrorists, however, it is also possible that attacks in the future may involve chemical and biological agents. It would be irresponsible to ignore this possibility and governments need to plan for such an eventuality. Emergency and other professionals involved in dealing with the aftermath of the release of a chemical or a biological agent face many challenges including detecting what has been released, determining who might have been exposed and what has been contaminated as well as how to clean up afterwards. All these activities require the use of equipment, measurement approaches and protocols which the scientific community could help develop. A UK Royal Society working group was convened in 2003-4 to consider these issues and this paper discusses its recommendations.

1. Introduction

The attacks on the twin towers in New York in September 2001, bomb attacks on embassies in Nairobi and Dar-es-Salaam, and the explosions in Istanbul and Madrid highlight the increasing threat from terrorism. Most recent attacks have used explosives to kill and injure. However, this is not the only threat.

In 1994 the Aum Shinrikyo sect attempted to spread anthrax on an unsuspecting population in Japan but failed, resorting in the end to the release of the nerve gas sarin in the Tokyo underground. Twelve people were killed by the sarin and upwards of one thousand injured. In the United States in 2001 finely milled anthrax spores were sent through the U.S. postal system to a number of prominent media representatives and members of the U.S. Senate. The resulting infections caused eleven cases of inhalation anthrax, five of them fatal, and eleven suspected cases of non-fatal cutaneous (skin) anthrax. Twenty three buildings were contaminated with anthrax spores of which three remain contaminated, and closed. The total cost of the decontamination exercise so far is some $800 million.

These recent incidents involving the use of anthrax give more credence to the argument that there may be further terrorist incidents involving the use of either chemical or biological agents. There is no shortage of substances to choose from for such an attack. Faced with such a threat government agencies and local authorities responsible for dealing with the consequences are confronted with huge problems.

Some of these are issues where scientists can help with advice and research. Two areas cry out for attention by the scientific community. The first is the need to improve methods of detection. Methods are required to detect a wide range of both chemical and biological agents. A second problem on which scientists can help is that of decontamination. Robust procedures for cleaning up individuals, buildings and contaminated land are needed urgently. To help address these issues the Royal Society convened a working group in January 2003 and its report, *Making the UK Safer: Detecting and Decontaminating Chemical and Biological Agents* was published in April 2004 [1]. The following sections are drawn from that report.

2. The Royal Society Report

2.1 Background to the Project

The Royal Society has a long-standing commitment to reducing the threat of biological weapons, and has produced two previous reports on this subject [2, 3]. The 2004 report *Making the UK Safer: Detecting and Decontaminating Chemical and Biological Agents* concerned the malicious use of chemical and biological agents against civilian targets such as key landmarks, transport hubs, postal sorting offices, government offices, water and power plants, or large gatherings of people. It examined how such attacks can be detected, how the agent(s) can be identified and quantified, and how targeted people and infrastructure can be decontaminated after attack. It concentrated on the implications for humans and their environment.

We did not address all potential terrorist threats. In particular, we did not deal with radiological threats (or "dirty bombs") nor infectious animal or plant diseases [4]. These are important issues and merit separate study.

For the purposes of the report, "detection" was used to cover systems and methods for early warning alarms, for monitoring and for identification and quantification.

There are many similarities in dealing with the consequences of a malevolent and an accidental release of a chemical or biological agent. Consequently, many countermeasures will be equally applicable in preparing the country against either type of incident.

The report was aimed principally at three groups: national and local government policy-makers involved in long-term planning to increase preparedness against a possible incident; emergency service staff (or "first responders") who would be directly involved in dealing with the consequences of an incident; and scientists and engineers working in areas that could be applied to extending existing detection and decontamination capabilities, particularly those who are currently unaware of the potential of their work.

Reducing the threat from chemical and biological agents requires political, economic, organisational and technological actions. A number of these issues were addressed by the House of Commons Science and Technology Select Committee in their 2003 report on the scientific response to terrorism [5]. A review article published in 2003 looked at the current U.S. situation regarding technology challenges [6]. The Royal Society report concentrated on where science, engineering and technology could help in diminishing the consequences of incidents by reducing vulnerabilities, improving the response of society by consequence management, modelling and the early warning of potential threats. Detection and decontamination are central to all of these issues.

An earlier report from the Royal Society entitled *Measures for Controlling the Threat from Biological Weapons* concluded that the scale of effectiveness of biological weapons against human populations in war and by terrorist attack had mercifully not been proven in practice and that, while it would be irresponsible to be complacent about the possible effects, it would also seem prudent not to overestimate them. That report also concluded that the main negative effect of a biological weapons incident might be panic and disruption of civilian services. Whilst the political and security scenario has changed since the 2000 report was produced, its conclusions remain valid [3].

Detection is an increasingly demanding and rapidly developing field, with considerable effort being devoted to new technical developments by both industry and the academic community. Detectors of various types are used throughout science and many could be adapted or developed for the specific or generic detection of chemical or biological agents. Issues which need to be considered are the priorities, concepts of use and implementation of detection systems as well as procedures for sampling. Who needs detectors and in what form is a key question.

Decontamination can be divided into the decontamination of people and the decontamination of structures including buildings, furniture, vehicles and equipment. The processes required for effective decontamination following an incident are not fully understood. This was clearly illustrated by the clean-up of the Senate buildings following the U.S. anthrax letters in Autumn 2001 and the difficulties experienced in restoring them to use. Science, engineering and technology can assist with decontamination in a wide variety of ways.

The consequences of either a chemical or biological incident can be greatly reduced if the agent can be rapidly detected; allowing appropriate countermeasures to be put in place as soon as possible. After an incident people need to know when it is safe to return and science, engineering and technology can determine when the environment is safe enough to justify a return to normal use.

Coordination and organisation of the research, development and planning relating to countermeasures against chemical and biological agents is fundamental to both reduce waste and concentrate effort in the most profitable areas.

Many scientific disciplines must be brought to bear upon detection and decontamination if effective detectors and procedures for decontamination are to be achieved. Key disciplines include microbiology, surface science, physics, chemistry, medicine and engineering. Novel science, engineering and technology have a vital role to play in meeting detection and decontamination challenges. The use of mathematical modelling of chemical and biological dispersions is also required to track plume movements. Scientific uncertainty, scientific advice and decision support for risk management must also be considered to help guide those who will make recommendations about areas being deemed safe for people to return.

Such developments could well occur in research disciplines that have not traditionally been focused on military or security related concerns, so it is vital to alert academics that their research might be relevant, even if there is no obvious link. Consequently, the coordination, commissioning and direction of relevant research are extremely important and are discussed throughout the report.

2.2 Conduct of the Royal Society Project

A working group chaired by Professor Herbert Huppert FRS prepared the report which the Council of the Royal Society endorsed. We asked key government departments and end-users of existing detection and decontamination technologies for their views on where science, engineering and technology could improve their existing systems. Based on the detailed information received, we issued a public call for evidence in May 2003. This was principally directed at scientists and engineers in the academic community, industry and government, and was aimed at determining where the cutting edge science, engineering and technology in the most appropriate areas exists and how it might be practically applied in the future. We also met a number of experts to discuss issues in greater depth. This extensive information forms the basis of the report.

3. Defining the Challenge

3.1 Types and Properties of Possible Agents

The approaches required to detect and decontaminate different agents will vary according to the properties of the agent in question. Potential chemical and biological agents have a range of physical properties and levels of toxicity. The physical form and properties of the individual agent will determine the most likely route of exposure to a material. Examples illustrating the range of chemical and biological agents are given in Table 1 and their different physical forms are outlined in Table 2.

Table 1. Range and Examples of Chemical and Biological Agents

Biological agents	Naturally occurring toxins	Synthetic chemicals
• Bacteria e.g. *Bacillus anthracis* (anthrax), *Yersinia pestis* (plague) • Viruses e.g. smallpox • Rickettsiae – *Coxiella burnetii* (Q fever) • Fungi – *Histoplasma capsulatum* • Modified bacteria and viruses	• Bacterial toxins including botulinum toxin • Naturally occurring bio-regulators • Ricin and related protein toxins • Mycotoxins (T2), aflatoxin • Palytoxin, batrachotoxin, tetraodotoxin, saxitoxin • Animal, plant, marine, snake, frog, toad, spider and scorpion toxins • Immuno-modulators, mood modifiers, analgesics, psychopeptides	• Chemical warfare agents, including nerve gases (mustard, sarin and VX), blister, blood and choking agents • Toxic industrial chemicals such as chlorine and phosgene • Highly potent pharmaceuticals and agrochemicals

Table 2. Physical Form of Examples of Chemical and Biological Agents

Physical form at room temperature	Example agent	Comments
Gas	Ammonia Chlorine	
Volatile liquid	Sarin Tabun	
Persistent liquid	Mustard gas VX Soman (when thickened)	Liquids can be thickened with polymers to increase their persistence
Liquid droplets	Viruses e.g. Variola (smallpox)	
Solid	Ricin Anthrax (spores)	Possibly in the form of spores, a dust or an aerosol

The level of toxicity will influence the requirements for detection and decontamination. For example, the level of decontamination required for a highly toxic agent is much greater than that for a less toxic agent. Also, it becomes more important to be able to detect small quantities of highly toxic materials because a minute quantity could have a serious impact on human health. For the most toxic materials, uptake of sub-milligram quantities per person could be lethal, but for the majority of chemicals higher doses are required. A particular toxic material will affect different species to varying degrees. It might thus be misleading to extrapolate lethal doses established in the laboratory to determine the number of people a small quantity of a toxin will kill. Also, there are fortunately considerable difficulties that would need to be overcome to successfully disseminate chemical and biological agents.

Materials can be inhaled, absorbed through the skin or ingested along with contaminated food or drink. Their toxicity will be altered according to the different routes of exposure. Although materials can be almost as toxic when inhaled as when injected, absorption through the skin usually leads to a reduction in potency. For example, anthrax is considerably more toxic when inhaled than when it is in contact with the skin.

The fate of agents following an incident will also alter the potential danger they pose. How long airborne agents stay suspended in the air and how air currents distribute them will determine their impact on humans. What happens to an agent when it lands on a surface, whether it becomes adsorbed or degrades over time, will also alter its potential effect on people. Physical form also affects detection, sampling and decontamination procedures. For example, detecting a highly toxic particulate will require sampling large volumes of air or large areas of a surface.

Detection, identification, monitoring and sampling procedures will depend on the physical form of the agent. For gases, liquids with a high vapour pressure or those in aerosols, detection will involve air sampling. In contrast, solid agents on surfaces will need to be removed from surfaces by swabs or wipes. Differences in the form of the agent have an impact on both detection systems and on sampling strategies.

3.2 Types of Possible Incident

There are many possible incidents that could occur, ranging in size from small to large. These could target air, food, water, people, livestock, horticulture, aquaculture, crop plants, strategic facilities or symbolic targets. The agents involved could be used in combination with radiological, nuclear or conventional explosives. The combination of these different factors will influence how the incident will best be dealt with.

4. Main Conclusions and Recommendations

The following sections set out the main conclusions and recommendations of the Royal Society report *Making the UK Safer.* The report contains recommendations regarding organisation and procedure; detection; decontamination; medical issues related to a chemical or biological incident; and the role of mathematical modelling

4.1 Introduction

Science, engineering and technology will be able to reduce the threat from chemical and biological agents. How this is coordinated and organised is crucial to the successful utilisation of developments in detection, decontamination, sampling and risk communication. Without political will and cost-effective implementation, organisational and technological innovation cannot deliver their full potential to make the UK safer.

It is currently not practical to constantly monitor the entire civilian population against unannounced chemical or biological attack. The difficulties preventing such an approach include the vast amounts of data it would generate, which could not be practicably analysed, and the logistical difficulties in setting up a network of appropriate monitors. Realistically, only a limited number of target locations can be monitored.

There is considerable expertise in dealing with chemical and biological agents in military scenarios. Whilst there are many differences between potential military and civilian incidents, some of the extensive military knowledge could be translated to a civilian context.

4.2 Organisation and Procedure

R1 The UK Government should establish a new centre to coordinate and direct the work required to improve the UK's capability and to minimise the impact of any civilian chemical or biological incident.

R2 The centre's main functions would be as follows.

- Determine, commission and direct the work required on planning, preparedness, research and development related to detection and decontamination.
- Assess and disseminate protocols and procedures for detection, sampling and decontamination.
- Evaluate detection and decontamination equipment and establish agreed industrial standards.
- Ensure information is shared effectively between different government departments and agencies, the academic community, industry and other interested parties, including the public.
- Establish the maximum levels of agents below which it is appropriate to permit a return to normal use following an incident.
- Work with the academic community, industry and the Research Councils where appropriate and seek to make full use of developments and potential funding in the U.S., Europe and elsewhere.
- Provide a clearly identified source of expert advice regarding chemical or biological incidents for government departments and agencies, first responders, NHS Trusts and national and local emergency planners.

It has become clear that the current system does not utilise the extensive expertise in local and national government, first responders, the academic community, industry and others to the greatest benefit to the UK. In order to harness that expertise, we recommend that such a centre has a significant, ring-fenced budget to commission work to develop and evaluate detection and decontamination equipment. The budget required by the centre will depend on the timescale envisaged for the work to be done and on the sophistication of the equipment and materials it aims to produce. Based on information from a number of sources, we estimate that a reasonable figure would be of the order of £20 million per year. It is important that the centre works with the Research Councils to identify promising research relevant to detection and decontamination. An outward looking approach would be essential for the success of such a centre, as would collaborations with other countries. If such a centre were not established then it would be extremely difficult for the appropriate parts of Government to be sufficiently aware of the diverse expertise to allocate this funding. Without this budget the UK will not gain the benefit of the existing but widely dispersed expertise.

The centre would bring together the existing expertise by working with the Defence Science and Technology Laboratory (DSTL), Health Protection Agency (HPA), Home Office, Department for the Environment Food and Rural Affairs (DEFRA), Environment Agency, Cabinet Office, Department of Health, Office of the Deputy Prime Minister, Department for Transport, the Research Councils, Office of Science and Technology, Department of Trade and Industry, National Health Service, first responders, the academic community and industry. A panel of expert, independent scientists should advise the centre, helping to guide the programmes and provide external scrutiny. The membership of such a group could be drawn from existing panels, and the Royal Society would be prepared to suggest potential additional members.

The centre should be located to maximise the overlaps with key partners. For example, the centre will need laboratories for handling highly toxic materials and it would be preferable for the centre to use existing facilities, such as Porton Down, rather than construct new ones.

The centre should also address the pressing need for investigation of the ways scientific uncertainties and technical difficulties impinge on all issues in responding to a chemical or biological incident, and how they are formalised for rational decision support at all stages. Dialogue between scientists, psychologists, politicians and the general public should be encouraged to improve the communication and public understanding of hazard and risk issues in relation to terrorist incidents, and any insights should be proactively incorporated into decision support. The efforts of the National Steering Committee on Warning and Informing the Public should be utilised in these challenges.

R3 We recommend that the next edition of the Cabinet Office document *Dealing with Disaster* [7], which is expected to pay more attention to dealing with chemical and biological incidents, clearly spells out the concepts of use for detection systems, so equipment and procedures can be designed and implemented accordingly.

In particular, this updated document should cover the scope for better coordination of pre-event action plans, scientific responses at the time of an incident, and timely implementation of scientific advances.

R4 Realistic exercises should be undertaken involving first responders, emergency planners and some civilians in order to test and develop the correct reactions to an incident.

In addition to providing a considerable measure of reassurance to the public, such exercises would be an integral part of staff training and preparedness. The proposed centre would advise first responders and emergency planners in running such exercises.

4.3 Detection

R5 We recommend that future work on detection systems should be concentrated on three objectives:

- Exploit new and existing science, engineering and technology for robust detection of chemical and biological agents.
- Develop point detectors for use by first responders at the scene of a suspected incident.
- Establish what information on background interferences and natural variability of agent levels might increase the reliability and sensitivity of different detection systems and decision making. Where appropriate the relevant data should be collected.

R6 The work on detection would best be coordinated and directed by the proposed new centre. If the proposed centre is not established then we recommend that an appropriate government department such as the Home Office Chemical, Biological, Radiological and Nuclear (CBRN) Team take the lead.

Considerable research is being undertaken on fledgling technologies by parts of the academic community that have little or no experience of working on military or security related projects. It is essential that this research be allowed to develop so that new potential technologies can be assessed, particularly in cross-disciplinary areas. This should include networking and data fusion activities and instrument development activities focused on solving real problems in a real environment.

One of the most urgent needs is for point detectors for first responders. We recommend that this should be a top priority of the research and development programme. Mobile or hand-held instruments for use by first responders at the scene of an incident would be extremely advantageous. The ideal instrument would be remote to avoid contamination of emergency staff.

Sampling methodologies must be consistent with detection methods and take into account the need for evidence gathering. The logging and storage of samples must be appropriate for the nature of the agent to be detected and will require liaison with the testing laboratories.

4.4 Decontamination

R7 With respect to decontamination studies, we recommend the following four priorities:

- Undertake a detailed review of the various options for the decontamination of people, buildings, vehicles and the wider environment.
- Assess the efficacy of decontamination procedures and technologies.
- Assess contact hazards from contaminated surfaces.
- Develop and implement techniques for avoiding secondary contamination in hospitals and ambulances.

R8 We recommend work on decontamination be coordinated and directed by the new centre, working in collaboration with relevant UK industries where appropriate. If the proposed centre is not established then we recommend that an appropriate Government department such as the Home Office CBRN Team take the lead.

This is an area where collaboration with appropriate UK industries would be extremely beneficial. For example, there is considerable expertise amongst detergent manufacturers. Human surface decontamination is still rudimentary: clothes are bagged and plenty of soap and water applied. More research is needed to determine the best technologies for generic cleansing of skin. There needs to be an extensive programme first to bring together existing knowledge and second to measure the effectiveness of decontamination formulations against a range of toxic chemical and biological materials. There is a need for rapid environmental decontamination at the incident site, using generic methods because the identity of the agent is usually unknown. Robust monitoring methods are also required to determine when the environment is acceptable for reuse.

4.5 Medical Issues Relating to Detection and Decontamination

R9 The occurrence of a chemical or biological incident may first become apparent through those affected reporting medical symptoms. Such reporting can therefore play a crucial role in the detection and subsequent decontamination of chemical and biological agents. We therefore recommend:

- Increasing training of clinicians in CBRN-related subjects by Medical Schools, led by the General Medical Council, to improve recognition of the relevant symptoms in individuals.
- Using medical intelligence analysis, in conjunction with the Health Protection Agency (HPA) and NHS, to improve recognition of a chemical or biological event at the level of the population and thus strengthen the resilience and the effectiveness of responses.
- Establishing systems for the long-term follow-up of exposed populations, with the Department of Health.

R10 These recommendations would best be integrated with the work undertaken by the proposed new centre.

Evidence based diagnostic techniques and appropriate training of medical personnel are important for detecting chemical or biological exposure. It is also vital to validate treatments for the effective consequence management of adverse effects on health. The treatment of unusual casualties will require special training for initial medical responders in the community, hospitals and health protection teams. Increased training should be extended to undergraduate and postgraduate medical training so that all doctors are aware of relevant toxicological and infectious diseases. This training should be delivered in a systematic approach and available to all institutions through electronic as well as traditional teaching methods.

There should be improved coordination of national and local electronic health surveillance systems to detect clusters of illness/symptoms and unusual diseases. Surveillance data should be real time and the HPA will have a key role in this development. The HPA and the NHS should utilise medical intelligence as it has the potential to make significant contributions to resilience and the effectiveness of responses. We welcome the announcement that the HPA is planning to establish a medical intelligence unit [8].

There are few data on the long-term risks to health in populations exposed to chemical agents. Agents such as mustard gas are suspected carcinogens, but the long-term effects of organophosphates are still unclear. Consequently, it will be important that exposed populations are identified and subjected to close long-term clinical follow-up.

4.6 Mathematical Modelling

R11 The proposed new centre should assess the current and future capabilities of mathematical modelling to provide real-time information to inform first responders and emergency planners.

The types of models that should be assessed include those to determine the extent of the initial contamination and potential re-dispersion, chemical plume dispersal for a range of built-up and open environments, predict the effectiveness of cleanup strategies and identify the potential impact on the civilian population. These models need to yield results in as near to real time as possible and should be tested against on-the-ground measurement of chemicals and simulates to validate and improve them.

Note

Alastair Hay was a member of the Royal Society Working Group.

References

[1] Royal Society (UK). Making the UK Safer: Detecting and Decontaminating Chemical and Biological Agents. London: Royal Society; 2004.
[2] Royal Society (UK). Scientific Aspects of Control of Biological Weapons. London: Royal Society; 1994.
[3] Royal Society (UK). Measures for Controlling the Threat from Biological Weapons. London: Royal Society; 2000.
[4] Royal Society (UK). Infectious Diseases in Livestock. London: Royal Society; 2002.
[5] House of Commons (UK). The Scientific Response to Terrorism. Science & Technology Select Committee, Eighth Report of Session 2002-03 HC 415. London: The Stationary Office; 2003.
[6] Fitch JP, Raber E, Imbro DR. Technology challenges in responding to biological or chemical attacks in the civilian sector. Science 2003; 302: 1350-1354.
[7] Cabinet Office (UK). Dealing with Disaster, Revised Third Edition. London: Cabinet Office; 2003.
[8] Home Office (UK). The Government Reply to the Eighth Report from the House of Commons Science and Technology Select Committee, HC 415-1 Session 2002-03. London: The Stationery Office; 2004.

Science and Technology Policies for the Anti-Terrorism Era
Andrew D. James (Ed.)
IOS Press, 2006

Cleanup after a CBRN Terrorist Event: What Do Users Need from the Science Community?

Konstantin VOLCHEK and Merv FINGAS
Environment Canada,
335 River Road, Ottawa, Ontario K1A 0H3, Canada

Abstract. The effectiveness of decontamination after a chemical, biological or radiological terrorist attack is determined by the cleanup technologies that are employed. This Chapter describes what is expected from the science community to develop adequate, reliable and economic decontamination methods and equipment. The needs for improved detection methods and adequate personal protection are also discussed.

1. Introduction

The risk of use of chemical, biological and radiological/nuclear (CBRN) agents in terrorist acts is now well recognised. Prevention measures would greatly reduce this risk but could never completely eliminate it. As a result, serious measures must be in place to respond to terrorist attacks that involve those agents. This includes the cleanup of affected buildings, machinery, land, water, and air [1, 2].

Immediately after an attack, first responders, including fire fighters, police and paramedics, would arrive at a site. They would carry out the decontamination of casualties. First responders are not expected however to do decontamination of buildings, equipment or land. This would be done by specially trained and equipped cleanup teams that are normally private sector contractors.

The cleanup after a terrorist attack would have much in common with the cleanup of contaminated sites or dealing with the consequences of industrial accidents. In fact, most of the potential chemical terror agents are represented by toxic industrial chemicals (TICs) or toxic industrial materials (TIMs) [3, 4]. Potential radiological agents are those radioisotopes used in either healthcare or industry [5, 6]. As far as biological agents are concerned, the consequences of their use in terrorist attacks and the response required will be somewhat similar to those of epidemics.

Using analogies between the so-called conventional cleanup and that after a terrorist attack are very useful in planning and executing the decontamination after a CBRN event. Both scenarios require the assessment of the severity of contamination: types and concentrations of the contaminants, effect on humans and the environment, exposure risks and so forth. Both require decisions on how the consequences of the contamination should be mitigated and in particular:

- Is cleanup required and to what extent?
- What methods and equipment should be employed?
- When will the affected area be considered safe?

- What to do with the decontamination waste?

"Conventional" cleanup methods and technologies could sometimes be used to deal with the consequences of CBRN terrorism. However, there are also major differences between the two scenarios. Those include:

- An urban environment with a high population density will be the likely target of terrorists, as opposed to industrial areas that are typical for "conventional" pollution. Cleanup operations would be likely to affect tens of thousands of people and interfere with such vital activities as energy and water supply, transportation and communications.
- Environmental remediation is normally focused on contaminated land or water. Contrary to this, the focus of CBRN cleanup efforts will likely be on residential, commercial and government buildings. This may involve the decontamination of materials such as drywall, ceiling tiles, upholstery and so forth. Many of these materials are notably porous and this factor is likely to reduce the effectiveness of decontamination.
- It is likely that a combination of different groups of agents, chemical and radiological, for example, will be used at the same time. This may require a multi-stage cleanup and/or may render some technologies unacceptable which otherwise would have been effective for a certain group of agents only.

In both scenarios, a balance must be found and maintained between the desire to clean "as much as possible" and the cost of the cleanup which may be extremely high. Health and environmental risk assessment is the tool that helps find this balance. This assessment however is normally a time consuming undertaking. It may not therefore be applicable in the case of CBRN terrorist attacks when a rapid response is vital.

The effectiveness of a cleanup, whether it is a conventional remediation exercise or a response to terrorist attack, will ultimately depend on the state of the art of the technology that is employed [7, 8, 9]. There is a strong need, driven by the remediation market, to develop more effective and economical methods of conventional decontamination. This need is even stronger in the case of CBRN counter-terrorism. This is due to the recognition that the risk of CBRN terrorism is imminent and its consequences may be devastating. This need to develop better tools and methods of decontamination sends an important message and provides incentives to the science and technology community who work in this field.

This Chapter will discuss the principal needs of organisations that will be engaged in the cleanup process and their expectations of the science community. The focus of this discussion is on actual decontamination technologies. Other related issues such as improved detection tools and adequate personal protection are also addressed.

2. Becoming Better Informed

First responders will provide an assessment of the situation following the terrorist attack. This will involve the detection of possible chemical, biological and radiological agents. The information collected will be crucial to planning and providing an immediate response that involves rescue, dealing with casualties and evacuation. It will also be used to decide whether decontamination may be necessary.

Once a decision is made that cleanup is required, it is likely that it will be necessary to gather additional information on the exact nature of the agents used in the attack, their quantities and concentrations, and specific areas or parts of a building that were affected.

All of this is necessary to assess what decontamination technologies should be used, where they should be used and what level of effort would be required. It is extremely important to the success of the cleanup that this information is reliable, accurate and readily available.

2.1 Knowing What to Expect

Given the fact that the choice of potential CBRN terrorism weapons is very broad, especially for chemical agents, it is difficult to plan agent-specific countermeasures. To address this challenge, two approaches might be used [4]:

- Groups of similar agents must be identified whose likelihood of use is high. For those groups of agents, specific decontamination procedures and appropriate equipment must be put in place. The likelihood must be defined by several factors including accessibility, technical feasibility of use and punitive damage. The limitation of this approach is that the likelihood can only be estimated to a certain extent. There is always a chance that unexpected agents would be used for which existing specific methods would be ineffective.

- Generic decontamination procedures and equipment must be developed and made available that would be effective on a large number of different agents, including those from different classes: chemical, biological, or radiological. The drawback of this approach is in the fact that different agents have completely different physicochemical characteristics. Accordingly, any generic decontamination method would work better on some agents and worse on the others.

Neither of these approaches alone would therefore provide an ideal solution to the problem. In fact, a combination of both seems to be the most productive. Users should expect therefore that they will be told what agents to expect and that they have technologies whose effectiveness is good enough to cover a broad range of possible agents.

2.2 Using Better Detectors

There are many systems on the market that can be used to accurately detect a number of potential CBRN terror weapons [7]. The task of detection is relatively easy in the case of radiological agents; however, it is difficult in the case of chemical agents. There are hundreds if not thousands of TICs and TIMs that could be used as terror weapons. They have different physicochemical characteristics and often require completely different detection approaches. Commercially available systems are normally designed to deal with one and sometimes a few types of agents. The challenge here is to have detection instruments that are capable of analysing a broad variety of agents. Another challenge is to have instruments that are sensitive to low concentrations of target agents but can operate without major interference from other, non-hazardous substances, which may be present at the site. For example, it is known that surfactants that are commonly found in today's environment may interfere with many of the potential agents. The science community must work to address these challenges and come up with advanced detection instruments and methods.

2.3 Knowing the Cleanup Targets

Removing 90 percent rather than 99.99 percent of the original contamination will often mean using completely different cleanup strategies and associated costs. A more rigorous clean is quite obviously normally more costly and it is important to know therefore how clean is "safe" [10].

In some cases, the existing guidelines for TICs and TIMs in the air can be used to assess the target level of residual concentrations of these substances after cleanup. This issue is quite complex, however, as some of these substances may be non-volatile. Their concentrations in the air may be low but they may still be present in large quantities in some areas that were not subjected to decontamination or where it was not effective enough. Consequently they can re-contaminate the building over a long period.

When a porous surface is affected it may virtually completely adsorb the agent so that there will be little evidence of its presence. The agent may remain entrapped for a long time and will render no harm. On the other hand, it may be released later as a result of increased temperature or humidity, and become very harmful.

There is relatively little information available that could be used to predict the behaviour of potential CBRN terrorism agents. Having this information would be very important in defining realistic and adequate decontamination targets. The science community could contribute to this area by providing such information, perhaps through a series of bench-scale tests. There may be instances when the entrapped agents will be found completely harmless; however, public perception will be strongly against leaving them at the site. In those cases, clear and science-based arguments from the science community could be helpful in reassuring the affected community.

3. Using Appropriate Cleanup Technologies and Methods

Similar to detection systems, there are many commercially available decontamination systems and related technologies that could effectively deal with chemical, biological, or radiological decontamination. It is extremely important that effective and cost-efficient technology and equipment are used.

3.1 Comparing Methods

The analysis of information available on different decontamination technologies suggests that there is no single procedure that has been used to evaluate the effectiveness of these technologies. In fact, very different test methods have been used in different cases so that it is quite difficult to compare results. For example, tests have been carried out using different agents, concentrations, and test matrices.

Being able to compare different technologies would help identify what is the most appropriate in a specific case. It is highly desirable therefore that the science and technology community develop a series of standard procedures to evaluate decontamination technologies. These procedures can be used for both existing and emerging technologies [12]. It is realised that there is no single procedure that will be applicable to all types of decontamination technologies. Instead several types of standard procedures may be proposed for the groups of technologies that use similar principles.

3.2 Focusing on Promising Methods

Although not everyone may agree, decontamination methods and equipment can be divided into three major classes, based on the nature of the decontamination process [4]:

- Decontamination using mechanical means;
- Decontamination using physicochemical means; and
- Decontamination using biological means, including natural degradation.

It should be emphasised that different types of processes are sometimes used in the same technology, such as in surfactant washing where mechanical removal with water is often accompanied by chemical reactions of alkali hydrolysis. In addition, natural degradation is often the result of not only biological but also physicochemical processes, such as photo degradation. Even though the above classification has its limitations, it may be helpful in identifying technology benefits, limitations and trends. In order to be considered for a cleanup, the technology must meet the following criteria:

- Effective removal or destruction of contaminants. The technology must be capable of achieving decontamination targets.
- Rapid rate of removal or destruction. A fast cleanup will be a major requirement, especially in the urban environment.
- Generation of no or little waste. If large volumes of waste are generated, they should be easy to treat.
- Acceptable cost. The use of expensive equipment or reagents will increase the cost. The above factors (effectiveness, removal rate, and waste generation) are also major cost factors.

These criteria serve as strong incentives for the development of new decontamination processes and improving the performance of existing technologies. Short summaries of trends and expectations for each of the technology groups will now be presented and reflect the authors' views based on their own experience and knowledge of technology developments in this field.

3.2.1 Decontamination using Mechanical Means

This group includes processes from water hosing to wiping to sandblasting. Since no biological or chemical destruction is involved, the effectiveness solely depends on mechanical removal. Understandably this group of technologies is most commonly used in radiological decontamination where the destruction of radionuclides is impossible and therefore not expected. It is also used, although to a lesser extent, in chemical and biological decontamination. Mechanical processes are relatively fast although their rate may be very slow when the upper layer of the surface needs to be removed to achieve the decontamination effect.

In many cases mechanical processes do not achieve an acceptable decontamination in one pass so that the treatment needs to be repeated several times. This results in increased time and cost of treatment. Another alternative is to intensify the processes. For example, high-pressure washing is capable of removing contaminants with its efficiency increasing with the increased pressure of water. At a very high pressure, however, the process becomes costly due to increased cost of equipment and energy consumption.

There have been reports of the use of pulse jet technology where water is applied under a varying pressure [11]. At its peak the pressure can be as high as 5000 bar which is much higher that any "conventional" high-pressure washing can achieve. Consequently the effect of removal is higher than in the conventional process.

Sandblasting is a common technology that provides good removal efficiency at a relatively high speed [8]. The disadvantages are the generation of large volumes of contaminated solid waste (sand plus contaminated materials) and the creation of contaminated suspended particles that can be easily inhaled.

The generation of large volumes of wastewater is a common feature of those mechanical methods using water. Sometimes the water can be regenerated and reused but this requires specialised equipment.

In view of the above technical challenges the use of cryogenic blasting provides a valuable alternative [9]. This method employs the bombardment of the contaminated surface with solid particles of carbon dioxide. The particles knock off the contaminants and then sublime without producing either liquid or solid waste. The only negative environmental impact is therefore the emission of carbon dioxide into the atmosphere.

Pulse jet technology and cryogenic blasting are only two examples of promising mechanical processes and they illustrate how scientific and engineering innovations can help overcome existing technology limitations.

3.2.2 Decontamination using Physicochemical Means

This is the largest group of technologies and includes a variety of processes, such as thermal and chemical oxidation, volatilization, hydrolysis and chelation. The majority of these technologies use water. In most cases, a complete or partial destruction of target agents is achieved on the surface to be decontaminated. In some cases, no further treatment of liquid wastes generated in the cleanup is necessary; although normally some sort of waste collection and treatment is required. Common limitations of this group of technologies include:

- Water may not adhere well to some surfaces, especially to vertical ones. Accordingly, there may be insufficient contact between the contaminated surface and active ingredients of the cleaning solution to achieve a desired level of decontamination.
- Water-based cleaning solutions are ineffective on hydrophobic substances such as many of the potential chemical terrorism agents.
- In order to improve the interaction between contaminants and active ingredients, large volumes of cleaning solutions are often used. This leads to larger volumes of liquid wastes that require disposal or further treatment.

In response to these challenges, cleaning formulations have been developed that are used in the form of foams or gels. They better adhere to surfaces including hydrophobic ones and enable the prolonged contact time required for a more complete decontamination. A number of formulations have been developed and commercialised. These include Sandia© foam [12], CASCAD© decontamination formulations [13], and L-gel© [14]. This is an area of on-going research effort that aims to develop better performing formulations and equipment.

Oxidation is arguably the most common process used in chemical decontamination. Oxidation can be greatly enhanced by introducing catalysts in the system. Fenton's process is a good example of a catalytic oxidation system with a salt of divalent iron serving as a catalyst and hydrogen peroxide serving as an active ingredient. This process is very efficient and inexpensive but its use is limited to aqueous systems only. It does not work on hydrophobic contaminants including those listed as potential chemical terrorism agents. Fenton's process can however be modified by incorporating either the peroxide or the catalysts into organic molecules to make them accessible to chemical agents. Research efforts are required to develop a modified Fenton's process that could be used in decontamination [4].

3.2.3 Decontamination using Biological Means

Biological processes, such as those employing bacteria or enzymes, have rather limited application. Their main advantages are their low cost and low environmental impact. These processes are typically much slower than mechanical or chemical processes and this seriously limits their potential application. They are utilised in relatively rare cases when rapid decontamination is not a major requirement [4].

As far as research needs are concerned, any development that could significantly increase the rate of biodegradation by an order of magnitude would create a revolution in the decontamination business. This goal may be unrealistic but any acceleration in biodegradation rate would make biological processes more competitive with alternative technologies. Users look to the science community to develop new types of bacteria and enzymes that can made decontamination faster.

3.3 Dealing with the Waste

Liquid, solid and sometimes gaseous wastes are generated in a majority of currently used decontamination processes. In some cases, such as radiological decontamination, the formation of wastes cannot be avoided. It is important however that the volume of waste is minimised. The cost of waste disposal or treatment can be comparable to or even exceed the cost of decontamination.

Waste minimisation can be achieved by employing technologies that generate less waste or, if those cannot be used, by reducing the volume of the generated waste. This waste volume reduction approach would not only make the final disposal or treatment less expensive but might also regenerate valuable products such as water that could be reused in cleanup operations.

In view of this, scientific and technological attention needs to turn towards the development of more robust mobile waste treatment/volume reduction systems that could be easily and effectively operated in a variety of cleanup cases. These systems might employ filtration, membrane separation, or adsorption technologies. They might also incorporate some treatment processes, such as advanced oxidation, so that the waste is not only reduced in volume but its hazard is also reduced [4]. There are already a number of mobile systems that are used for solid and liquid waste treatment. The challenge to the science and technology community is to adapt them to the variety of wastes that could be produced in cases of CBRN decontamination.

4. Improving Protection of Personnel

CBRN decontamination teams may operate in conditions that are extremely hazardous. Take pesticides as an example. These are relatively non-hazardous in a liquid form so long as there is no direct skin contact or ingestion. On the other hand, if pesticides are used in terrorist acts they may be dispersed and form aerosols. In this form, their toxic effect could be much stronger. Effective personal protection for members of decontamination teams is therefore vital. Once again, the science and technology community can make an important contribution by developing adequate personal protection for use in each case of decontamination, in view of the agents used, their speciation, and concentration.

5. Improving Training

It is reasonable to expect that CBRN decontamination teams will have substantial experience in dealing with chemical, biological or radiological substances. In fact, many of these teams will have previous experience in environmental cleanup or building decommissioning. It is realised however that specific approaches may be very different in the case of a CBRN cleanup operation. This is due to the nature of the agents, their dispersal mode, and the type of surfaces to be decontaminated.

It is important therefore that the teams receive adequate training before they engage in an actual cleanup [10]. The science community must provide expert advice on how this training exercise should be organised. What surrogate agents should be used (and at what concentration) to mimic "real" CBRN agents? What level of personal protection should be in place? What technology and equipment should be utilised in exercises? What waste minimisation, treatment and disposal issues may arise?

6. How can the Science Community Help?

The needs of first responders dealing with CBRN contamination are not limited to just those discussed in this Chapter. For example, reliable models are sought that can describe and predict the behaviour of CBRN agents and better assess the associated risks. There are many other areas where research has to be done to facilitate a more effective cleanup and it is unlikely that those directly responsible for cleanup after a CBRN incident will be able to solve scientific problems on their own. This work can be done only by specialised research organisations such as universities, government laboratories or private sector engineering firms doing the research. On the other hand, the research organisations are expected to work in a close contact with first responders to get a necessary feedback. First responders must advise the scientists on where their research efforts should be focused.

7. Benefits to Industry at Large

Decontamination after CBRN terrorist attacks should be considered a part of the civil defence industry. Its main goal is protect society against man-made intentional disasters. On the other hand, we discussed in the introduction section of this Chapter that CBRN decontamination strategies and approaches have much in common with those used in site remediation. It is quite reasonable to expect that new and effective CBRN decontamination technologies, which are being developed to serve the needs of civil defence, will also find an application in future site remediation projects. One should also expect that those technologies may find niches in dealing with a variety of non-terrorist man-made and natural disasters and other cases when fast, effective and cost-efficient decontamination is necessary. Examples include accidents at chemical or nuclear plants but may also include outbreaks of infectious diseases.

8. Conclusions

While it appears impossible to completely eliminate the risk of CBRN terrorism, at least now, it is possible and absolutely necessary to be prepared for it. This includes the ability to employ adequate detection methods, use modern and effective decontamination technologies and equipment, deal efficiently with decontamination wastes, and do all of this in a safe manner. The challenge for the research community is to generate new technological solutions that will better equip and protect decontamination teams.

Acknowledgements

The information about the needs and expectations of CBRN decontamination technology users was gathered and analysed during a study supported by the Chemical, Biological, Radiological or Nuclear Research and Technology Initiative (CRTI), Canadian Department of National Defence, Project Chapter CRTI-02-0067RD. The authors appreciate valuable comments and recommendations provided by Norman Yanofsky of Defence Research and Development Canada, Laura Cochrane of Vanguard Response Systems Ltd. and Patrick Lambert of Environment Canada.

References

[1] The Home Office (UK). The Release of Chemical, Biological, Radiological or Nuclear (CBRN) Substances: Guidance for Local Authorities. London: The Home Office; 2003. Available at: http://www.ukresilience.info/cbrn/cbrn_guidance.pdf.

[2] Lawson JR, Jarboe TL. Aid for decontamination of fire and rescue service protective clothing and equipment after chemical, biological, and radiological exposures. NIST Special Publication 981. Washington, DC: National Institute of Standards and Technology, U.S. Department of Commerce; 2002.

[3] Purver R. Chemical and biological terrorism: the threat according to the open literature. Ottawa, (Ontario): Canadian Security Intelligence Service; Updated 2000. Available at: http:// www.csis-scrs.gc.ca/ eng/miscdocs/tabintr_e.html#toc.

[4] Fingas M, Volchek K, Hornof M, Boudreau L, Punt M, Payette P, Best M, Wagener S, Bertrand K, Cousins T, and Haslip D. A project to develop restoration methods for buildings and facilities after a terrorist attack. In: Proceedings of the 27th AMOP technical seminar; 2004 June 8-10; Edmonton: Environment Canada; 2004. p 453-76.

[5] Schopfer C. Radiological preparedness case studies: training for accidents and dirty bombs. The New Jersey Centre for Public Health Preparedness; 2003. Available at: http://www.njcphp.org/resources/ACOEM.html.

[6] Waller E, Volchek K, Cole D. Technical Aspects of Radiological Terrorism. Draft Report W7714-2-0619. Ottawa: Defence Research and Development Canada; 2003.

[7] Fatah AA, Barrett JA, Arcliesi RD, Ewing KJ, Lattin CH, Helinski MS, Baig IA. Guide for the selection of chemical and biological decontamination equipment for emergency first responders. NIJ Guide 103-00. Washington DC: U.S. Department of Justice; 2001. Volume I (http://www.ncjrs.org/pdffiles1/nij/189724.pdf) and Volume 2 (http://www.ncjrs.org/ pdffiles1/nij/189725.pdf).

[8] Manion WJ. Review of key decontamination and dismantlement technologies. In: LeSage LG, Sarkisov AA, editors. Nuclear Submarine Decommissioning and Related Problems. Dordrecht: Kluwer Academic Publishers; 1996.

[9] Argyle MD, Demmer R, Archibald K, Tripp J. Chemical and non-chemical decontamination development at the Idaho Nuclear Technology and Engineering Centre. In: Proceedings of the 7th International Conference on Radioactive Waste Management and Environmental Remediation: Nagoya, Japan; 1999.

[10] The Royal Society (UK). Making the UK Safer: Detecting and Decontaminating Chemical and Biological Agents. London: The Royal Society; 2004. Available at: http://www.royalsoc.ac.uk/ files/statfiles/document-257.pdf.

[11] Vijay MM. Properties and parameters of water jets. In: Proceedings of the International Conference on Geomatics 91; Ostrava (Czech Republic); 1991; p 207-22.

[12] Sandia National Laboratories. Sandia decon formulations for mitigation and decontamination of CBW agents. Albuquerque (NM): Sandia National Laboratories; 2004. Available at: http://www.sandia.gov/SandiaDecon/demos/ demos.htm.

[13] Vanguard Response Systems. CASCAD Decontaminating Chemicals. Stoney Creek (Ontario): Vanguard Response Systems; 2004. Available at: http://www.vanguardresponse.com/ products_cascad_3. shtml.

[14] Heller A. L-gel decontaminates better than bleach. Lawrence Livermore National Laboratory; 2002. Available at: http://www.llnl.gov/str/March02/Raber.html.

Part 3

Public Policy Responses

Andrew D. James (Ed.)
IOS Press, 2006

A Framework for Homeland Security Research and Development: The United States' Perspective

Dr. Penrose C. ALBRIGHT
Assistant Secretary
Science and Technology Directorate
U.S. Department of Homeland Security

Dr. Holly A. DOCKERY
Special Assistant for International Policy
Science and Technology Directorate
U.S. Department of Homeland Security

Abstract. The developed nations of the world hold an asymmetric advantage in the fight against terrorism - an advantage due to a well-funded and highly capable science and technology enterprise. This enterprise has been deployed over the past 50 years in part to support the military establishment, with incredible success. However, the basic way of thinking about military science and technology does not hold for the civil operational environment, and hence it should not be assumed in general that military technologies can simply be transferred to the homeland security community. In fact, that rarely if ever happens successfully. Furthermore, there are substantial policy constraints and issues surrounding the development of a homeland security research and development community. These issues led to the formation of a separate and dedicated homeland security research and development capability. A specific and critical issue associated with homeland security, and its research and development capability, surrounds the use of weapons of mass effect. Understanding the risks and prioritising among these threats brings additional difficult issues to the table that may never be resolvable. In part because of these very difficult policy dilemmas, the new Department of Homeland Security (DHS) was created, with a Science and Technology Directorate that is on equal status with the other operational agencies within the Department. This status reflects the President's vision for employing the advantage held in science and technology against the threat of catastrophic terrorism.

1. Introduction

In the war against terrorism, the United States and its allies enjoy an asymmetric advantage in science and technology. It is important to press home this advantage with an international research and development enterprise for homeland security comparable in emphasis and scope to that which has supported the military community for over fifty years. This is appropriate, given the scale of the mission, and the catastrophic potential of the threat. Such an effort in science and technology applied to homeland security should be aimed at both evolutionary improvements to current capabilities and the development of revolutionary new capabilities.

The enabling technologies developed within this enterprise should have several attributes. Importantly, these technologies must not only make us safer, but also make our daily lives better; while protecting against the rare event, they should enhance the commonplace. Thus, the technologies developed for homeland security should fit well within our physical and economic infrastructure, and our national habits. Many of the envisioned systems will be national, continental, or even global in scope, and thus the technologies must scale appropriately, in terms of complexity, cost, operation, and sustainability. In many cases the operators will be local police, firemen, or even volunteers, so the technology must require minimal training and maintenance. System performance must balance the risks associated with the threat against the impact of false alarms and impediments to our way of life.

Research and development aimed at homeland security should not be an open-ended process leading to ensembles of incoherent projects, but rather the result of a constant examination of the vulnerabilities, constant testing of our security systems, and a constant evaluation of the threat and its weaknesses. In short, the approach to research and development should be strategic in nature. The development process must be disciplined, and thus, despite perceived urgency, unacceptable technical or programmatic risk must not be accepted in favour of overly aggressive schedules; history has shown repeatedly that such an approach results in failure. Technologies and systems must be tested and demonstrated throughout the development process, and risks retired in an orderly manner. Pilot programmes should be developed as appropriate prior to large-scale deployment, and guidance from first responders and operators aggressively sought.

Homeland security as defined in the United States is far broader than countering terrorism. It also includes maintaining and enforcing national sovereignty along our borders, dealing with natural and environmental disasters, and other conventional missions. However, there can be no doubt that the events of September 11th, 2001, the anthrax letters that appeared soon afterwards, and, importantly, an examination of the motivations of our enemies, has led to a new and intense focus on domestic security, and served as the impetus for the formation within the United States of the new Department of Homeland Security (DHS).

In this Chapter we will first discuss some of the conceptual issues associated with this new emphasis on homeland security, and with the new Department. We will then discuss the role of the science and technology community in homeland security, and the history of the Science and Technology Directorate in the new Department. The Chapter concludes with a brief discussion of the successes to date, and some of the challenges.

2. Conceptual Shifts Associated with Homeland Security Technology Development

2.1 Addressing Weapons of Mass Effect

Perhaps the greatest issue surrounding contemporary counterterrorism is the potential for truly catastrophic terrorism - biological, nuclear, radiological, and chemical. In part this concern is associated with the motivations of the adversary, which are in large part to inflict ever greater harm in terms of numbers of civilian casualties or economic damage. Terrorist organisations have made clear their desire to acquire such weapons, and at least for biological, radiological, and chemical weapons, the technological infrastructure and expertise needed for their production does not represent a significant barrier. Furthermore, the critical ability to produce fissile material is no longer in the hands of a few countries, and is spreading. A large-scale biological attack, or the explosion of a nuclear weapon on domestic soil would be epochal in its effect. Prior acts of terrorism revolved around the use of explosives on aircraft or in public spaces, with casualties ranging from a few to (for a small number of aviation-based acts of terrorism) a few hundred, and in some cases the intent of the act was limited - the terrorists might notify officials prior to the explosion in order to clear people from the area. The attacks of September 11th killed over 3000, and the use of weaponised anthrax in an aerosolised release (as opposed to mailing it in letters), or a nuclear weapon, could kill ten times or a hundred times more. The scale of this threat and the softness of the targets require new ways of thinking.

Public health officials have had to come to terms with the idea that someone might deliberately introduce, for example, smallpox into the population with the simple desire to kill a large number of people. Even if convinced of the real possibility of a deliberate attack, the initial assumption has been that conventional public health methods would be effective in controlling the outbreak. A substantial analytic effort was required to demonstrate that such a threat could not be dealt with using the paradigms that the public health community has used to eliminate natural contagions. Similarly, the notion that a deliberate introduction of foot and mouth disease into the livestock population can simply be dealt with in the same manner as a natural outbreak has clearly been shown to be wrong. In both of these examples the scale of a deliberate attack in terms of size, timing, and spatial diversity, would quickly cause traditional methods to fail. Other examples are equally easy to demonstrate; the means for dealing with a radiological attack on a downtown business district are different to those associated with an incident involving a nuclear power plant or environmental cleanup at a research facility.

2.2 Differences in Technology Needs for Military versus Civil Security Missions

Historically, countermeasures to chemical, biological, radiological, and nuclear (CBRN) threats have been developed within the military science and technology establishments to deal with the potential use of these weapons on the battlefield. However, while the science and technology base generated in those efforts is invaluable, the technology simply cannot be transferred to the civil community. One obvious reason for this is that militaries must train and equip themselves for deployed operations, and hence assume an extensive and constant supply chain, depot and spares infrastructure, as well as the availability of a cadre of specialists trained to service the equipment. The civil community, in general, cannot support that same infrastructure or a dedicated trained workforce. It is also unlikely to devote considerable resources to countermeasures that may only be needed once in a lifetime. Furthermore, military equipment is usually designed to be used episodically, with brief periods of intense use followed by intervals of intense maintenance and replenishment. In the civil environment, the equipment in question needs to always be operational - 24 hours a day, 7 days a week, 365 days a year, year in and year out. Additionally, many of the underlying assumptions associated with the performance requirements for technology developed for military use are predicated on its use by soldiers of a certain age and level of physical fitness, and with the acceptance of a particular level of personal risk. The civilian population that must be protected includes people of all age groups, in varying degrees of health, and who enjoy a strong legal framework to minimise risk to the individual. Clearly, military technology must be reengineered to address total cost of ownership and use within the civilian environment if it is to be deployed for homeland security purposes.

A further, perhaps more subtle, issue with the use of military technology is that there can be a significant reliance on secrecy. In part this is due to the episodic nature of its use. Keeping some of the characteristics of military equipment closely held is both practical and desirable. However, this will be rarely the case in civilian deployments. As mentioned above, constant round the clock, and generally widespread, deployment makes it nearly impossible to protect the characteristics of the equipment. Furthermore, civilian decision makers who must trust the equipment are generally unwilling to do so without having deep insight into its operation. For example, public health officials will be unlikely to declare a major public health emergency on the basis of a sensor reading from a "black box". Finally, measures to maintain secrecy often impose undue cost on the operation of the technology. Hence, technologies intended to counter the terrorist threat, particularly the CBRN threat, will perhaps have their characteristics and detailed specifications held "sensitive" (i.e. not readily accessible to the general public) but will in general not be protected in the same manner as are military technologies. One consequence of this is that it will be nearly impossible to simply transfer equipment from the military to the civilian community. A further consequence is that it must be assumed that the threat knows the characteristics of the available countermeasures, and thus the performance of these systems must either be robust to that knowledge or the relevant characteristics must be easily and repeatedly adapted in order to render that knowledge continually obsolete.

2.3 Providing Appropriate S&T Support to All Levels of Government

Homeland security is an "all-government" issue. Within the United States that means the involvement of federal, state, local, and tribal agencies. In most cases the implementation of countermeasures occurs at the community level, and hence the role of the federal government is to provide technical assistance and advice, to develop standards, and provide funding. Of specific interest here is the fact that hardly any local jurisdiction within the United States maintains any organic capacity for providing scientific assistance. Decisions related to the purchasing of equipment rely either on vendor claims or on whatever advice may be sought from or provided by the federal government. While this is an issue for all sorts of equipment, such as interoperable radios or self-contained breathing equipment, it is exacerbated by the much more technical nature of the CBRN threat and associated countermeasures. This is in distinct contrast to the military, intelligence, and nuclear weapon establishments (known colloquially as the "national security" community) where the efforts are "owned" by the federal government and where each agency maintains a robust and active scientific activity. Even within the national government, most agencies involved with homeland security, such as customs and immigration services, or law enforcement, have themselves rarely possessed or seen the need for a robust research and development capacity. Again, in general, the relevant capability exists within the national security establishment, and while in some cases the civil community may benefit from "hand me down" technology, as we have indicated above successful transfer of relevant capabilities will be rare.

2.4 Developing a Meaningful Expression of Risk

A further issue is that it is nearly impossible to quantify the risk associated with the CBRN threat. Hence, persuading agencies or private firms to acquire technology for dealing with that threat can be difficult. While it is important to deal with the more conventional threats such as bombings and hostage-taking - extraordinarily daunting in an era of suicide bombers and a willingness, as demonstrated by the Beslan school massacre in Russia, to inflict harm on the most helpless - the CBRN threat presents nearly unimaginable consequences. However, the classic risk calculus does not apply. Risk is normally thought of as the product of probability of occurrence multiplied by the consequences. However one interprets probability of occurrence (e.g. some measure of the degree of difficulty for the threat to mount the attack), in some cases that "measure of belief" about the threat will be judged to be extremely low (e.g. the nuclear threat), while the consequences extremely high. In effect, one is forming a judgment based (qualitatively) on multiplying zero times infinity.

Ultimately, those decisions will instead devolve to how a decision maker "feels" about the risk from the threat, and on the expected costs and efficacy of the investment decision. While it can be argued that eventually all major investment decisions come to this sort of judgment, the lack of a robust analytic risk methodology for these threats is problematic. One approach is to use scenario-based risk assessments. This at least allows illumination of the consequences of an attack and can also play an important role in informing decisions regarding where in the chain of events investments are most likely to be effective - intelligence, detection, response and recovery and so forth. However, scenario-based tools obviously beg the question of the likelihood assigned to the scenario, and on whether the marginal investment of a dollar should go to countering the nuclear threat, chemical threats, or explosive detection technologies. Should analytic intelligence be available (e.g. a laboratory found in a cave in Afghanistan) then that certainly would colour judgments made about dealing with that threat, but absent such serendipitous evidence prioritised decisions must instead be made based on the need to anticipate such high consequence threats and ultimately on the need for capabilities to address them. The extraordinarily difficult problem of setting with any analytic rigour relative priorities for countering the various threats of interest (indeed, argued here as essentially impossible) has been mitigated within the United States by the high priority placed by the President and Congress on providing adequate resources for addressing all these threats (at least to the extent that solutions can in fact be developed, and thus the resources spent wisely).

It is worth noting that uncertainties associated with determining risk with any analytic rigour also significantly impact the judgments made by insurance firms. In part the inability to render actuarial analyses, coupled by the potentially very large consequences (and hence claims), has made it often impossible for technology developers to deploy new capabilities because insurance is not available. Thus, a further policy issue that must be dealt with if technology is to be leveraged for homeland security is how to manage the risk and liabilities faced by firms developing solutions to the threats of interest. Clearly, the government needs to balance the need to see technology deployed for protecting the public against the public's right to redress should technology fail.

3. Elevating the Role of Technology Development in the Department of Homeland Security

The Department of Homeland Security (DHS) was created to accomplish several goals. First, there was a need to unify various agencies operating at our borders. Bringing together the customs, immigration control, border patrol, and transportation security elements into a single agency assures unity of mission and deconfliction of resources. Bringing the Federal Emergency Management Agency into the Department allows appropriate prioritisation among federal resources devoted to prevention, protection, response and recovery, and unifies planning for both catastrophic natural and manmade disasters.

However, there were two wholly new activities created. First, there was a need within the U.S. government to unify in an analytic enterprise foreign intelligence data with domestic law enforcement data, and to further merge that information with a continual assessment of our vulnerabilities. Hence, a new enterprise was created (the Information Analysis and Infrastructure Protection Directorate), with an official responsible for this function, and a concomitant budget. A further new capability was created with the formation of the Science and Technology Directorate.

A decision could certainly have been reached to include homeland security, and particularly the science and technology component, within the broad mission space associated with the Defense Department. Much of the existing research and development capacity existed there. However, we have already discussed the issues associated with using technologies developed for military purposes in the civil sector. Serving that sector implies an entirely different mindset when developing requirements and making systems trades. More subtly, though, within the overall mission of defeating terrorism, there will be a constant tension between the allocation of resources to interdict terrorism overseas by destroying infrastructure, and killing and capturing personnel; and for protecting against terrorism at home and mitigating the consequences of an attack should it occur. The tendency of the military will be to serve the needs of the war-fighter first, and assign a much lower priority to serving the specialised needs of the civilian users. This circumstance would not allow us to construct the appropriate technology systems to provide the multiple layers of defence necessary to defeat terrorism.

To ensure that all aspects of the war against terrorism receive adequate resources and attention, efforts to defeat terrorism overseas will continue to be advocated, planned, programmed, and funded by the Departments of Defense and State. However, it was decided that domestic efforts to combat terrorism should reside elsewhere. That is, the competition for resources between overseas and domestic efforts aimed against terrorism and its effects should be resolved at the Presidential level, and not within the processes of the Department of Defense, for example.

3.1 Rationale for the Science and Technology Directorate in DHS

One crucial issue was that of which functions should (or should not) be incorporated within the new Department. Ultimately, day-to-day management (and allocation of budget) requires a line organisation to take responsibility for the function. Unification of most functions associated with homeland security into a single department allowed concentration of effort and resources on the mission. However, in the broadest sense "homeland security" is a central function of any national government. It encompasses activities in virtually every federal agency. For example, public health is a national enterprise in its own right aside from the counterterrorism mission. So, although a new focus on the public health aspects of terrorism was necessary, a judgment was made that due to the overall size, scope, and capability of public health as an extant government enterprise that function should not be brought in to the new Department. Indeed, it is worth noting that the Department of Homeland Security's budget for countering bio-terrorism is roughly $400 million, while that within the National Institutes of Health (NIH) and the Center for Disease Control and Prevention (CDC) is roughly $1,800 million. DHS funding is focused primarily on environmental and facilities monitoring, technical intelligence and threat assessment, and cleanup technologies while the NIH and CDC focus on developing medical countermeasures. As the Department was being created judgments of this nature were made affecting activities across the government. The net effect is that, despite the existence of a department concentrated on the homeland security mission, there will always need to be a body that reports to the President and guides policy development and coordinates related programmes across the entire federal government.

This Chapter has already noted that few resources had been dedicated to performing research and development for the purposes of homeland security, and very few of the agencies brought in to the new Department came with organic scientific capacity. It is interesting to note, however, that the original language in the President's bill for the new Department did not refer to a Science and Technology Directorate. Rather, the original title of this portion of the Department's structure was called the Chemical, Biological, Radiological, and Nuclear Directorate (or CBRN Directorate) and this was to have responsibility for "securing the people, infrastructures, property, resources, and systems in the United States from acts of terrorism involving chemical, biological, radiological, and nuclear weapons or other emerging threats". However, the intrinsic technical nature of that mission also led to a further responsibility: "conducting a national scientific research and development program to support the mission of the Department..." [1]. However, when presented with the bill, Congress believed insufficient emphasis was being placed on the underlying science and technology mission, and thus the Directorate was renamed. However, the counter-CBRN mission remained.

What is clearly important, though, is a recognition by the nation's leadership of the importance of science and technology to domestic security, particularly given the technical nature of the new threat environment. Furthermore, the application of science and technology to homeland security, given the historic lack of attention to this issue presents the potential for significant high payoff opportunities with sustained focus of effort. Perhaps most importantly, as we have noted, the approach to domestic security requires a different systems approach that allows for the relevant operational environment, decision-making structure, and resources. The integration of missions from various agencies into the new Department also presented an opportunity to provide an integrated technological focus that would support the entire mission of the Department. These technologies must address terrorism, and particularly the inherently technical issues surrounding the CBRN threat, but also support the full set of responsibilities of the other components within the Department. Travel and trade must be made more secure, but also more efficient. Technologies developed for responding and recovering from disasters should be interoperable and address the spectrum of hazards for which the Department is responsible. Since the capabilities to be developed will in many cases be acquired and used at the non-federal level, the technology should be multipurpose where possible. Adequately dealing with these needs requires a dedicated science and technology enterprise that addresses the entire mission space of the Department with an efficient and disciplined systems engineering approach.

3.2 Organisation of the Science and Technology Directorate

In order to provide a policy planning, programming, and budgeting framework for the new directorate, the research and development effort was organised around "portfolios". A policy planning staff was created for each area of activity within the Directorate, led by a portfolio manager. Each portfolio manager is responsible for developing a vision, and architecture, and an implementation strategy over the near, mid, and far term for their specific area of responsibility. The focus is on the operational end state and research and development activities are seen simply as a means for achieving that state. Thus, for example, portfolios were created for countering the chemical, biological, radiological/nuclear, and explosive threat. Each portfolio would create a vision that dealt with the entire mission space: anticipating the threat; protecting against the threat; and responding to and recovering from an attack should it occur. A key responsibility for each portfolio manager is balancing among these activities. In order to prioritise among needed capabilities, a gap analysis is performed that compares the risk from the threat (based on an understanding of the technical capability of the threat and the consequences of an attack) with currently available countermeasures, or countermeasures already under development.

Importantly, portfolios were also created to address the needs of the operational user. Thus, portfolios associated with the border agencies, with emergency management, with intelligence analysis, and with the state and local user communities were formed. The role of these portfolio managers is somewhat different, in that they are to understand the operational issues within their user community, develop technical or systems solutions to them, and provide technical advice. As an aside, it is also clearly important to have a capability for identifying cross-cutting issues among the portfolios, and taking advantage of synergies that might exist. This too was recognised and implemented within the new Directorate.

A further important capability was to create a standards activity that would develop and promulgate national performance standards and uniform test protocols for homeland security equipment. Standards not only allow informed procurement decisions by state and local agencies that otherwise have at best limited access to competent technical advice, but also provide guidance to developers on acceptable performance.

Finally, a statutory system of risk and liability management was created that limits the amount of insurance needed by firms seeking to deploy homeland security technologies. This requires a careful assessment of the efficacy of the technology or service provided, and the liability issues surrounding the company.

A significant function - arguably the most important function - for the Science and Technology Directorate is to mobilise the scientific community for the homeland security mission. As noted before, this community exists and has served the national security enterprise very well. It is comprised of private sector firms large and small, national laboratories, universities, and government labs of all sizes, and spans the space of scientific disciplines with extraordinary depth. However, as also noted above, the operational environments are substantially different than those previously served. Substantial interaction with the user, often a technically unsophisticated user, is required. The scientific community serving that user must approach the problems in terms of needed capabilities and operational imperatives.

4. Successes and Challenges

It is far too soon in the history of the new Department to talk about failures and it is nearly impossible to sort out issues in execution (due to the start-up nature of the activity) from true problems. However, some observations can be made, in no particular order of priority.

First, as noted above, homeland security responsibilities will always extend beyond the reach of a single department. Particularly where substantial successful activities already existed, and where substantial dual use (between counterterrorism and baseline missions) of the infrastructure and talent pool exists, those capabilities were not brought into the new Department. Thus, a formal coordination body must exist that cuts across federal agencies. This function is being performed by three primary components of the Executive Office of the President: the Homeland Security Council, the Office of Science and Technology Policy, and the Office of Management and Budget. While the system operates well, inevitably adjustments among departments and within these structures will continue to occur.

Second, although the events of the fall of 2001 have resulted in enormous attention and resources being placed on homeland security, it does not mean that there was no activity in that area prior to then. Certain agencies and structures had on a small scale evolved to fill obvious gaps, and had taken on responsibility (outside their baseline mission space) for aspects of homeland security. The advent of the Department of Homeland Security, and hence of an agency with domestic counterterrorism as its baseline mission, has meant that these "extracurricular" activities are no longer particularly relevant. This has led to some transient confusion of roles and responsibilities; transition of people, facilities and capabilities back to baseline missions; and to as yet unresolved duplication of effort. This issue will resolve itself over time as agencies reclaim budget for their baseline mission space and as the aforementioned coordination activities mature.

Third, in large part because of the lack of attention paid to research and development for homeland security technology in the past, there are obvious gaps addressable with generally short-term efforts. This is the so-called "low hanging fruit." For example, basic scientific knowledge developed within the national security establishment often needs reengineering to be suitable for the civilian operational environment, but such development generally does not require new discovery or invention. Unfortunately, there is a temptation to neglect the scientific base and long range research. Striking the right balance between the ratio of short-, mid- and long-range investigation now, and then over time evolving that mix represents a significant management challenge.

Fourth, and perhaps the most difficult, is developing a capability for prioritising among threats. As discussed above, this is difficult if not impossible across the CBRN spectrum, and also in determining the mix between resources devoted to these threats and to the more conventional needs of the Department.

Despite these daunting challenges, many successes have occurred. Environmental monitoring for biological pathogens has been established in over 30 cities in the United States and these extraordinarily sensitive systems have never had a false system alarm. The next generation system is undergoing operational field trials, and there is a long-range research programme aimed at developing "smoke alarm"-like devices. The aim is to dramatically reduce cost and maintenance requirements while keeping or improving detection performance. Chemical sensors have been prototyped for detecting release of nerve and other agents on subway platforms. The aim is to restrict access to the platform and stop trains should an attack occur, thus preventing small numbers of casualties from becoming far larger. Programmes in major facilities such as airports are in addition demonstrating the practicality of redirecting air flow and developing real time evacuation prompts to further reduce the effect of an attack. Radiation monitors are being deployed at our border ports of entry and the next generation of technology already being piloted. This new generation discriminates sources of radiation, thus allowing a priori sorting of naturally occurring sources from sources of interest, and hence reduction of thresholds while maintaining (and not saturating) secondary inspection capacity. These technologies are being deployed and new generations tested also at highway choke points (e.g. toll booths) and at air and sea cargo facilities. A major effort is underway to establish standards for interoperable communications equipment, with an initial focus on incident site interoperability, and an eventual goal of community-wide interoperable communications. Standards have already been issued for radiation detection equipment deployed for homeland security purposes, for personal protective equipment used by local public safety practitioners, and a new standard is imminent for test kits that determine the risk associated with an unidentified "white powder."

5. Conclusion

The previous section contains but a brief listing of successes. The new Department of Homeland Security's Science and Technology Directorate is a new start up, but with a budget of about $1.2 billion has the resources and capacity to assert the asymmetric advantage held by the developed world. The focus is not just on the science but the adaptation of that science to the civil operational environment. Given the global nature of the threat, the scope of work that needs to be done, and the need to assure that no developed country becomes a "soft target" for those threatening our common values and way of life, the effort must be international in scope, and strategic in its focus.

References

[1] A Bill to establish a Department of Homeland Security, and for other purposes. 2002. Available at: http://www.whitehouse.gov/deptofhomeland/bill/

Science and Technology Policies for the Anti-Terrorism Era
Andrew D. James (Ed.)
IOS Press, 2006

Development of a Science and Technology Response for CBRN Terrorism: The Canadian CBRN Research and Technology Initiative

Camille A. BOULET
CBRN Research and Technology Initiative, Defence R&D Canada
305 Rideau Street, Ottawa, Ontario, K1A 0K2, Canada

Abstract. In Canada, the CBRN Research & Technology Initiative (CRTI) represents the federal commitment to providing science and technology (S&T) solutions for national security and CBRN preparedness. It is a joint, interdepartmental initiative between 15 science based departments and agencies, security based departments and central agencies to strengthen Canada's preparedness for, prevention of, and response to a CBRN attack by fostering new investments in research and technology. Three significant innovations were built into the development and implementation of the CRTI. These are the use of a Consolidated Risk Assessment to assess CBRN events and derive S&T programme targets; the establishment of laboratory clusters to address capacity requirements for response; and the formulation of a project model that values collaboration and that is open to all sectors of the national innovation system (government, academic and industrial). Within Canada, these have changed how the federal S&T community responds to a national priority and represent a significant shift in roles for federal science and technology from supporting the development of government policy to a more active role leading the nation's innovation system to address a national priority. Underpinning each of these elements is a strong, community building approach to ensure the broadest possible response to the public security and safety S&T objective.

1. Introduction

The terrorism events and anthrax letter attacks in the fall of 2001 brought public safety and security and the need for Chemical, Biological, Radiological and Nuclear (CBRN) counter-terrorism preparedness into sharp focus. In Canada, the CBRN Research and Technology Initiative (CRTI) represents the federal commitment to providing science and technology (S&T) solutions for national security and CBRN preparedness.[1] It is a joint, interdepartmental initiative between 15 science based departments and agencies, security based departments and central agencies to strengthen Canada's preparedness for, prevention of, and response to a CBRN attack by fostering new investments in research and technology. Initially funded with a five year, $CAN.170 million budget, it specifically targets CBRN terrorism by addressing capacity, knowledge, science, and technology gaps.

Recent Canadian experiences with such pathogens as Severe Acute Respiratory Syndrome (SARS), Bovine Spongiform Encephalopathy (BSE or "mad cow disease"), and Avian Influenza (AI) have served as practical models to understand the emergency preparedness and response capabilities required to manage a CBRN event, how the science community supports the federal response, and to test the approach taken by the Canadian S&T community.

The purpose of this Chapter is not to provide a detailed review of the programme or the projects already underway in CRTI but to outline the Canadian experience in formulating a new model to engage the S&T community in response to a significant public safety and security challenge. This Chapter will describe the fundamental principles of the S&T programme that include a risk assessment methodology for prioritisation of investments, defining the S&T outcomes that contribute to the new security mission, developing project criteria that encourage collaboration, and providing examples of increased capacity and capability enabled by CRTI.

2. Engaging the Innovation System

2.1 Federal Innovation Networks of Excellence (FINE)

During 2000 and 2001, as part of the Canadian government's commitment to innovation, the federal S&T community was developing an initiative known as the Federal Innovation Networks of Excellence (FINE). FINE was designed to integrate federal, university, and private sector science capacity to produce solutions to emerging public good science issues as well as innovative products and services for economic development and national goals through federally led networks. By September 2001, the process had already identified CBRN protection, as well as other public safety and security areas such as disaster consequence management, information and physical infrastructure security, and the knowledge management of surveillance and intelligence information as areas where the FINE initiative could be employed.

In support of government deliberations to enhance public security and safety in advance of the fall 2001 budget, the federal S&T community was requested to examine the roles that it could play in supporting the nation's capacity and capability to address the CBRN threat. The Department of National Defence/Defence R&D Canada (DRDC) led the assessment, with support from client communities such as the Solicitor General and Office of Critical Infrastructure Protection and Emergency Preparedness.

The purpose of this assessment was to provide clarity as to where gaps existed in scientific capability and capacity and to suggest ways to address these shortfalls either through the clustering of capacity or through new investment to build both capacity and capability. Capability referred to the S&T knowledge and technology needed to fully address CBRN terrorism including threat assessment, detection and identification, treatments and therapies, remediation, containment and forensic investigation. A lack of a capability would identify a specific gap in knowledge, science or technology in the response to such an event. Capacity referred to the S&T resources, personnel, equipment or laboratories required to respond to a CBRN event. Capacity requirements were related to the magnitude of the impact and consequences of an event.

The departments and agencies involved in this process developed a consensus on the relative risks of the various possible events involving a specific hazard directed against a specific target. Risk was interpreted as the product of the probability of the occurrence of a terrorist attack multiplied by the magnitude of the consequences of the attack. The seven highest risk scenarios[2] selected for further assessment were:

- Biological terrorism against human populations in an urban setting;
- Chemical terrorism against human populations in an urban setting;

- Chemical terrorism against human populations using the food/water supply;
- Biological terrorism against human populations using the food/water supply;
- Biological terrorism attacks against agro-systems (animals) and the environment;
- Biological terrorism attacks against agro-systems (plants) and the environment; and
- Nuclear reactor event affecting human populations, agro-systems, and the environment.

2.2 S&T Capabilities and Capacity

Examining capability and capacity in the context of the functions that the S&T community would be called upon to perform developed an assessment of the ability of the national S&T structure to respond to each of the seven high-risk scenarios. The assessment took the form of a capability gap analysis but issues of capacity also were identified based on the following tasks:

- Criminal investigation (e.g. forensics);
- Crisis management and immediate reaction (e.g. hoax resolution, hazard detection and identification, protection of first responders and treatment of victims, and public communications); and
- Consequence management (e.g. containment, decontamination, casualty care, mitigation, forecasting and monitoring, and public confidence measures).

The general findings identified a number of areas where Canada's scientific capacity was insufficient to deal with a potential CBRN terrorist event. Some capacity shortfalls could be addressed by the clustering of federal and national laboratory resources without necessitating a new investment. This clustering approach could provide a surge capability adequate to meet the expected needs of several of the scenarios but would necessitate a formal approach to establish the clusters and identify the baseline capability and capacity of each cluster member. There were, however, other areas where even after available national capacity and capability was brought to bear, shortfalls would continue to exist.

As a result of this assessment, the following measures were recommended to improve the nation's preparedness to respond to terrorist attack:

- Establish an authority for the co-ordination of federal S&T in CBRN counter-terrorism and the funding for a federal laboratory response network;
- Clarify the roles and mandates of Federal Departments and Agencies as they pertain to the operational and S&T aspects of CBRN disaster response;
- Create clusters of federal laboratories as elements of a federal laboratory response network that will build S&T capacity to address the highest risk terrorist attack scenarios;
- Create a fund to build capability in critical areas, particularly those identified in the scenarios that address biological and radiological attack;
- Accelerate technology into the hands of the first responders community and other operational authorities; and
- Provide funds to those areas where national S&T capacity is deficient owing to obsolete equipment, dated facilities, or inadequate scientific teams.

In Canada, the CBRN Research and Technology Initiative (CRTI) represents the programme mandated to address the measures described above. It is a joint, interdepartmental initiative between Public Safety and Emergency Preparedness Canada, Health Canada, Atomic Energy of Canada, Canadian Nuclear Safety Commission, Environment Canada, Agriculture and Agri-Food Canada, Canada Food Inspection Agency, Department of Fisheries and Oceans, National Research Council, Natural Resources Canada, Royal Canadian Mounted Police, Canadian Security Intelligence Service, Treasury Board Secretariat, Privy Council Office, and Defence R&D Canada.

2.3 Roles for the S&T Community

Four roles were identified for the S&T community. These went beyond the obvious research and development and include most importantly areas where the S&T community's knowledge and experience can be applied to national priority areas such as security, safety, and counter-terrorism. These roles are:

- Scientific Analysis and Advice: provide scientific analysis and advice across the spectrum of strategy, policy, acquisition, and operations;
- Research Development Test and Evaluation (RDT&E): conduct RDT&E to close knowledge gaps, enable interoperability and reduce risk through conception and fielding of new and improved national capabilities;
- S&T Forecasting: Anticipate emerging public security threats to improve prevention and implement mitigation and countermeasure strategies;
- Outreach to Non-Federal Entities: Engage the full innovation system in the private, academic, government and end user communities to identify and provide effective S&T solutions.

These roles also indicate an important shift for federal science, a shift from the traditional role within the federal system where research supported the development of policy to one where federal science was now leading the response to a national security priority.

3. A Scenario-based CBRN Consolidated Risk Assessment

3.1 The CRA Methodology

To develop sound counter-terrorism prevention, preparedness, and response mechanisms, there was a need to develop and employ a systematic analysis of the risk of CBRN terrorism, the key knowledge and scientific gaps, and investment priorities. This systematic approach was also the basis for developing a shared understanding of the nature of CBRN terrorism amongst the various communities required for an effective response.

The CRTI developed a Consolidated Risk Assessment methodology (CRA) as the foundation for programme direction and prioritisation of S&T investments, projects, or activities. The results of this annual assessment are used to identify knowledge, S&T or key capability gaps and help operational communities and first responders to more effectively prepare, prevent, and respond to CBRN terrorist acts. It is particularly valuable as it provides a rigorous and defensible methodology to make decisions in a resource-constrained environment.

The basic elements of the CRA methodology include:

- Broad consultation with and participation by the S&T, operational and law enforcement, and intelligence communities;
- Development of characteristic scenarios that briefly describe the nature of an individual CBRN event;
- Evaluation of the public safety vulnerabilities reflected in each of these scenarios using a Vulnerability Matrix that considers Relative Technical Feasibility and the Impact that such a scenario would have if successful;
- Assessment of the degree of risk associated with each scenario using a Risk Matrix that factors Vulnerability and Intelligence Judgment;
- Systematic analysis of the CBRN hazards, targets and scenarios to identify Key Mitigating Factors where S&T or knowledge can be applied; and
- Systematic analysis of gaps to identify investment priorities that consider the full spectrum of prevention, preparedness and response requirements.

The CRA examines representative scenarios that describe how a particular hazard can be employed against a target. These scenarios cover a range of CBRN terrorism events including:

- CBRN attacks against people in cities and enclosed spaces;
- CBRN terrorism against agricultural systems (plant and animal);
- CBRN terrorism using food, water, or consumer products to affect human populations; and
- CBRN terrorism as it may affect critical infrastructure.

The first step of the CRA is to determine the Vulnerability by assessing the Relative Technical Feasibility and the Impact that a particular scenario could have if successfully executed. Technical Feasibility considers aspects of the availability of the biological material, its deployment and dissemination, any equipment necessary for production, and the technical expertise and knowledge required for planning and executing an attack. This leads to a Technical Feasibility rating for each scenario of high, medium, low, or very low where high would be a scenario that would be relatively easy to undertake and where very low would be a scenario requiring advanced knowledge or materials not readily available.

Impact is evaluated, considering human losses, intensity of response at the local, provincial or federal levels, overall disruption of capability and capacity, and economic losses. The Vulnerability rating is assigned based on the product of Impact and Relative Technical Feasibility. The analysis takes into account disproportionate impacts where, for example, the impact on local health care systems or other unintended consequences such as affects on public confidence can be much greater than anticipated based solely on the number of direct victims or financial loss.

Secondly, to assess the overall probability of a particular scenario, an Intelligence Judgment is provided for each scenario. The Intelligence Judgment considers parameters such as the amount and reliability of intelligence reporting, knowledge of terrorist activities, and terrorist intent and thus provides an assessment based on available information. Risk is finally assigned as the product of Vulnerability and Intelligence Judgment. In the context of the CRA, the Intelligence Judgment is considered to be the best measure of probability for a given scenario. The risk rating ranges from immediate through high, emerging and finally discretionary.

3.2 Investment Priority Areas

The assessment of each scenario in the CRA is a departure point for further analysis of the immediate and high-risk scenarios, to provide guidance for S&T projects and for evaluating and testing federal response measures. An analysis of key mitigating factors, capability gaps, and of S&T targets for research and development, as well as other activities that can mitigate the effects of an attack, is part of the CRA methodology. These capability areas consider the full spectrum of response requirements as well as broader issues such as collective Command, Control, Communications, Coordination and Information (C4I) for CBRN planning and response. Other S&T capability areas are:

- S&T for equipping and training first responders;
- Prevention, surveillance, and alert capabilities;
- Immediate reaction and near-term consequence management capabilities;
- Long-term consequence management capabilities; and
- Criminal and forensic investigation capabilities.

Two significant areas for S&T emerged from this analysis that to date have not formed visible parts of most CBRN response programmes. The first is the role that public confidence and psychosocial factors play in prevention, preparedness, and response to a terrorism event. The requirement for sound communication strategies that address public concerns as well as plans that account for public variability will be central to a federal government's strategy to successfully respond to any such event. This also reinforces the need to involve the broadest scientific expertise, particularly in less traditional areas of CBRN S&T such as the social sciences and humanities community. The second area, gaps in the knowledge base on the technical feasibility, impact and consequences of CBRN terrorism, as opposed to CBRN warfare, forms the basis for S&T investment in itself.

3.3 A Knowledge Gap: S&T Dimensions of Risk Assessment

An important example of a knowledge gap is the need to undertake quantitative assessments of the feasibility, impact and consequences of CBRN terrorism scenarios. Traditional hazard assessment of large scale, militarily significant CBRN attacks can provide some guidance. However, it is often of limited value when compared to actual CBRN terrorism events, consequences, and outcomes. Significant knowledge gaps, particularly in the areas of biological terrorism targeting animals and plants as well radiological terrorism, emerged.

As an example of the knowledge gaps in the S&T dimensions of risk assessment, the increasing number of letters alleged to contain anthrax being sent to health clinics, government offices and other locations in the U.S. and an alleged "anthrax letter" incident at a government office in Canada prompted a study of the hazard that such letters could produce. At that time, the "passive" form of dissemination of anthrax from opening a contaminated envelope was thought to only pose a direct hazard to the person opening the letter, some hazard to others in the immediate vicinity, and a minimal risk outside the area. These so-called expert conclusions derived from the knowledge of the impact and consequences of anthrax used in a militarily significant quantity.

In reality, a quantitative hazard assessment for anthrax letters, used as a terrorist weapon, had never been conducted. When a study was designed to specifically examine the hazard from such anthrax letters, using a small quantity of a biological simulant, it showed that the passive dissemination was considerably more dangerous and would have far more serious consequences than previously considered [2]. Opening an anthrax letter would readily produce a respirable aerosol and cause rapid and extensive contamination at hazardous levels. Overall, similar knowledge gaps continue to exist with respect to the vulnerabilities posed by CBRN agents in terrorism as opposed to their use in warfare and specific studies are needed to better understand the nature of these events.

3.4 Capability and Capacity Gaps: The Anthrax Letters

Several scenarios have been identified as posing an immediate or high risk to public safety or security. Most of these scenarios are chemical or biological attacks directly against human targets or the use of biological agents in agro-terrorism. One of these scenarios analysed in the CRA was a covert biological aerosol attack employing an anthrax spore-containing letter. This can be used to illustrate how the risk assessment process can be validated using actual experience of terrorism together with public heath data, to derive capacity and capability gaps that can be addressed through an S&T programme. In the CRA, a "covert aerosol attack using a non-contagious biological material" scenario is assessed as requiring a medium level of technical feasibility, having critical impact, and thus the vulnerability to such a scenario is high. The risk itself is assessed as immediate. An analysis of the capabilities and capacities needed to manage such a scenario, validated by actual events, allows for identification and prioritisation of S&T areas.

In October 2001, a series of anthrax attacks occurred in the United States. After an initial report of inhalation anthrax in Florida, several cases of inhalation and cutaneous anthrax occurred in 6 other states. These cases resulted from letters in which an anthrax powder was mailed to various media and government offices. A total of 22 cases of anthrax were identified, 11 cutaneous and 11 inhalation with 5 fatalities resulting from inhalation anthrax [3].

The U.S. anthrax attacks represented a further escalation in the illicit use of CBRN materials. An unknown person and/or group were able to acquire the necessary knowledge and capability to launch an attack using a biological warfare (BW) agent. These attacks identified new consequences, intended or unintended, from a BW agent that had not been identified as a result of previous chemical terrorism. The attacks demonstrated the tremendous impacts on both persons and infrastructure. 32,000 people were placed on an initial course of antimicrobial therapy of which over 10,000 were recommended to take a 60-day course of therapy because of factors indicating possible exposure. The anthrax letters further caused extensive contamination of facilities and infrastructure [4, 5]. Consequence management of the anthrax attacks extended for months and even years unlike the immediacy of a CW event [6].

In particular, the requirement for rapid diagnostics placed a tremendous burden on the public health system. There was a need to identify exposed persons to ensure that timely and appropriate medical treatment measures were provided [7]. Medical diagnostics, prophylaxis, and therapy emphasised the need for an immediate surge capacity not previously demonstrated in chemical terrorist attacks. The gap analysis and review of key mitigating factors for several biological scenarios in the CRA shows that the requirement for rapid diagnostics with high throughput capacity is a broad lesson and priority for new S&T investments.

The demonstrated impacts of the anthrax letter attacks can be measured in terms of the number of casualties, the gaps in the technical capabilities of first responders, the response capacity and capability burden it placed on all levels of government, and the clear demonstration of the value of using biological agents to achieve terrorist objectives.

The requirements for rapid diagnostics, surge capacity amongst response organisations, and non-linear consequences of an event were reinforced in recent Canadian experiences with SARS and AI. These were not terrorist attacks but SARS in particular demonstrated a greater than anticipated impact on the overall health care system that was disproportionate to the immediate consequences of the public health event.

4. Integration of Federal S&T to Address Capacity

4.1 Laboratory Clusters

Under the framework approved by the Treasury Board of Canada and to support the Government of Canada's response to terrorism, Canadian federal science based departments and agencies have been directed to develop a network of laboratories in three disciplines: chemical, biological, and radiological/nuclear. These networks are known as clusters and are comprised of federal laboratories from departments and agencies that have mandated roles and responsibilities in a terrorism event such as Health Canada, the Canadian Food Inspection Agency, the Royal Canadian Mounted Police, and Environment Canada. Other government departments or agencies also participate based on particular knowledge, expertise, or laboratory capacity role they have with respect to terrorism response and the particular CBRN hazard.

Each Laboratory Cluster is formally tasked to:

- Ensure its preparedness through the development and maintenance of a cluster operations plan;
- Develop the roles and procedures by which the laboratory cluster would support department operational mandates during a bioterrorism event;
- Develop appropriate working relationships between cluster members with particular emphasis on engaging first responders; and
- Manage cross-cluster interactions.

The specific activities related to CBRN terrorism managed by the laboratory clusters include:

- Threat assessment;
- Surveillance, alert, and warning;
- Crisis management and immediate reaction;
- Consequence management;
- Attribution and criminal investigation;
- Operational preparedness; and,
- Surge capacity and sustainability.

Through the CRTI, each cluster is provided funding and support to acquire technology and capacity to improve its response to a terrorism event. These clusters can also plan exercises, studies, or enter into formal collaborations to address a specific S&T project. Broad departmental participation in the laboratory clusters ensures that federal laboratory resources are available in most provinces and regions in Canada. This facilitates access to specialised facilities and expertise and can, for example, minimise transportation of sample requirements. The Canadian CRTI Biological Cluster serves as a model for the development of a multidisciplinary response network to support both national and international response to bioterrorism. It allows for access to equivalent facilities and trained personnel to provide surge capacity for diagnostic and environmental sampling.

A multidisciplinary approach is also required to address all phases of consequence management and post-incident investigations. The establishment of ongoing cooperation and collaboration helps ensure that capacity and capability are available, in advance, to manage the consequences of a bioterrorism event. The recent public health and animal disease outbreak events in Canada illustrate that the value of the cluster model goes well beyond bioterrorism to provide a response to broader public health incidents.

4.2 The Biological Cluster in Action

The laboratory cluster model has already been validated as a result of recent public health and animal disease outbreaks in Canada and the resulting engagement of the Biological Cluster and its members in the federal response. In 2003, Cluster members supported Health Canada in the investigation of the aerosol hazard from SARS in hospital environments and coordinated access to specialised equipment and surge capacity at CFIA's National Center for Foreign Animal Disease.

In 2004, following the outbreak of Avian Influenza in British Columbia, CRTI Cluster members were called on to assist CFIA operations. The Meteorological Service of Canada (MSC), the Canadian Meteorological Centre, Defence Research and Development Canada – Suffield and Agriculture and Agri-Food Canada participated in the response.

The MSC deployed an emergency surface weather station and an emergency upper air station to provide special weather forecasts as well as fine scale and atmospheric dispersion modelling. This equipment is housed at MSC regional centres across Canada to facilitate rapid deployment. The meteorological modelling supported the CFIA investigation of the source and origin of the Avian Influenza outbreak and Defence Research and Development – Suffield deployed aerosol-sampling equipment to track the virus. Agriculture and Agri-Food Canada assisted by providing geographic information system support and information, allowing CFIA to focus their efforts better on direct consequence management. The Canadian Meteorological Centre and the CFIA are also collaborating on the development of simulation programs on the spread of animal and zoonotic disease.

The laboratory cluster model allows for access to knowledge and expertise not readily available in any single department. The involvement of the cluster in supporting CFIA operations against Avian Influenza provided a good example of CRTI pan-cluster support, in terms of the deployment of specialised equipment and expertise, as well as an optimal use of leverage and resources within a system. The laboratory cluster experience has already demonstrated that ensuring the broadest possible participation amongst various organisations can ensure rapid access to knowledge, expertise, and capacities as vital components of a full response capability.

5. Formulating an S&T Programme to Value Collaboration

5.1 Key Principles

Four key principles were used in the formulation of the CRTI S&T programme in addition to the laboratory cluster model. These were designed to ensure that the CRTI, once implemented, would be able to value collaboration across the horizon of S&T performers, broadly engage the operational and response communities, and provide outcomes that met the roles identified above. These principles were:

- Single programme focused on a national security priority open to government, academic, and private sectors of the innovation system;
- Project funding requires collaboration amongst S&T performers;
- Project leadership at the federal S&T level; and
- Co-investment through contributions by project partners to lever existing programme funding.

There is no doubt that funding matters. By establishing mandatory criteria for project funding that reflected the desired outcome of increased collaboration and expanded investment, by definition the projects would address this key value. To further promote collaboration and engagement of the three pillars of the innovation system (government, private and academic sectors) the CRTI projects require that a federal laboratory retain the leadership for each project. This further reinforces the functional direction and leadership role that certain departments will have in events involving public health, animal disease outbreaks, and the criminal nature of terrorism by helping ensure that the knowledge and outcomes of the projects are retained in these departments.

5.2 CRTI Project Categories

Projects are selected via a competitive process and this represents another significant departure from traditional S&T programme models in particular for federal scientists. The selection process employs a quality and relevance review and the selection criteria reflect the specific outcomes expected for the project categories. Partnering and leveraging is core to all of the project categories. Each CRTI project comprises activities that are designed to achieve agreed objectives within a specified timeframe using assigned resources. Three project categories were developed:

- Technology Acquisition: These projects are intended to establish or enhance the infrastructure and equipment of the Laboratory Clusters that support the CBRN response. These acquisitions will typically be made in the year in which they are funded and will be "off-the-shelf" purchase of existing technology. The CRTI priority will be afforded to the most critical gaps in capacity, consistent with laboratory cluster roles and responsibilities;
- Technology Acceleration: These projects are intended to accelerate the commercialisation and transition to use by First Responders and other operational authorities of technologies that address key capacity gaps. These projects will typically be completed within six months to two years of funding approval, involve technology that is "in the pipeline", and must include at least one industrial partner;
- Research and Technology Development: These projects are intended to close knowledge and capability gaps in the S&T and operational communities so as to enable effective response to future CBRN threats. These projects encourage partnering and leveraging by requiring the collaboration of at least two federal partners in addition to any academic or industrial participants.

Technology Acceleration, Research & Technology Development, and Technology Demonstration projects are sought through a two-step process based on a broad solicitation from government, industry and academia. Initially minimal submissions sufficient to assess quality and relevance are screened and those accepted are then invited to submit full proposals. This two-step process provides the advantages of a streamlined submission process, facilitates collaboration, and provides equitable access to all sectors of the innovation system.

5.3 Initial Measures of S&T Project Success

In the first two phases of project selection, 41 projects in Technology Acceleration and Research and Technology Development have been approved. The total investment by CRTI in these projects is approximately $CAN 75 million whereas the total project investment, as a result of co-investment of the project partners, exceeds $CAN 157 million. This demonstrates the substantial commitment by the innovation system to the S&T objectives and programme. The partnerships formed through the project teams are also extensive. Over 10 federal departments, 22 private sector companies and 20 academic partners are participating in the CRTI project portfolio.

These projects address key capability or knowledge gaps in CBRN prevention, preparedness, and response. The overall understanding of the S&T gaps provided by the risk assessment process and the engagement of the S&T community has resulted in a comprehensive portfolio and coordinated projects that address the full response spectrum. For example, the radiological and nuclear project portfolio includes projects that address detection of radiological sources, wide area surveillance, biological dosimetry technologies to rapidly increase throughput of possible exposed individuals, as well as supporting environmental monitoring and meteorological modelling. Key knowledge gaps are also addressed through such projects as the comprehensive risk assessment of radiological dispersion devices (RDDs) that will examine all aspects of RDD production and use, including source acquisition, construction risks, delivery mechanisms, consequences of use, and possible countermeasures.

Operational community requirements such as improved materials, personal protection standards, and novel forensic methods have been supported. Another project will for the first time examine psychosocial issues of CBRN terrorism by providing an integrated psychosocial and bioenvironmental risk management framework for CBRN agents, and practical field-based training tools to enhance the capability of first responders to mitigate the psychosocial and human health impacts of CBRN threats and attacks. Clearly there is broad engagement of the innovation system in Canada.

6. Extending the Collaborative Model

This model has also contributed to the successful development of a collaborative programme with the United States to address public security science and technology. In December 2002, U.S. Secretary of Homeland Security Tom Ridge and Minister Manley of Canada agreed to expand the Canada-U.S. Smart Border Accord to include an element to address science and technology as it contributes to the nations' mutual border security. Consistent with the scope of the Accord itself, a broad and holistic view of border security and its S&T dimensions has been taken, one that is not uniquely focused on the security of a physical perimeter.

The Public Security Technical Program (PSTP) has been established with a mission to "Collaboratively deliver proactive S&T solutions that advance our national capabilities to prevent, respond to and recover from high-consequence Public Safety and Security events." Four mission areas have been identified for initial collaboration:

- CBRNE: the capability to prevent, prepare for and respond to CBRNE threats to public security, whether derived from terrorist or criminal activity, natural causes or accidents;
- Critical Infrastructure Protection (CIP): the capability to ensure the robustness, reliability and protection of physical and IT facilities, networks, services and assets, which if disrupted or destroyed would have a serious impact on the health, safety, security, economic well-being or effective functioning of the nation;
- Disruption and Interdiction (DI): the capability to identify and stop terrorists/criminals and their activities, including surveillance, monitoring, disruption and interdiction of their activities through intelligence, law enforcement and border and transportation security;
- Systems Integration, Standards and Analysis (SISA): the performance, integration and interoperability of national and international public security and emergency management capabilities and supporting systems, including the enabling standards, and vulnerability and systems analyses.

Central to the development of a strategic plan for the PSTP is the use of a coordinated risk assessment. This assessment is based on an agreed set of scenarios that will be evaluated for their impact, consequence and technical feasibility to assist in identifying capability gaps. The strategic plan will identify S&T priorities that will be the focus of cooperation on a time horizon of 3-5 years and a programme roadmap will describe specific joint projects and activities that will be pursued. The PSTP projects will demonstrate their value by providing scientific analysis and advice, conducting RDT&E to close knowledge gaps, anticipating emerging public security threats, and engaging the full innovation system in the private, academic, government, and end user communities to provide effective S&T solutions.

7. Conclusions

Three significant innovations were built into the development and implementation of the CBRN Research and Technology Initiative model. Within Canada, these have changed how the federal S&T community responds to a national priority and represent a significant shift in roles for federal science and technology from one of supporting the development of government policy to a more active role of leading the nation's innovation system to address a national priority.

The Consolidated Risk Assessment allowed science and technology, operational and intelligence communities to collaborate on targeted products that would serve as guidance for prevention, preparedness and response activities across government. It facilitated interaction and consensus building between scientists and the operational enforcement and intelligence communities. In Canada, this interaction has resulted in a framework that prioritises S&T investments for CBRN counter terrorism. Emergency preparedness planning and public health responses also benefit from this risk assessment based approach. The knowledge gained can be used to provide information to first responders and establish measures to address the risks through response planning, projects, partnerships and exercises.

The Laboratory Clusters have brought together multiple federal S&T departments and agencies to form non-traditional partnerships with a focus of improving capability and capacity to respond to a national priority. The cluster model demonstrated that there is substantial existing capacity within the federal system to support the response to CBRN events and further identified significant gaps that warranted new investments. The Biological Cluster has already been effectively engaged to assist in the response to SARS and AI outbreaks.

The CRTI project model, which values collaboration and allows equitable access to all sectors of the innovation system, has been developed and successfully employed. The CRTI represents one of the first instances where funding is not earmarked to specific S&T performers eschewing the traditional separation of funding to the academic community through grants, support to government researchers through departmental budgets, and industrial research programmes. In addition to the enhanced capacity and capabilities established with the federal science based departments and agencies, over $CAN 7.8 million in funding has been awarded to innovative projects in Canadian universities and close to $CAN 20 million in Canadian industry.

The CRTI model requires co-investment by project partners to enhance the programme through leverage but more importantly to ensure that projects address the needs of the participating community. This co-investment is a key indicator of the value of the project and is important in ensuring the project outcomes and deliverables will be successful. To date, the contributions in kind and direct investment by project partners have more than doubled the total CRTI portfolio value.

To address the full dimensions of prevention, preparedness, and response to CBRN terrorism events requires fully coordinated and integrated national programmes and even international programmes. Thus a systematic approach to understanding the risk and identifying capability gaps is an important basis for the development of international strategic science and technology investments. This systematic approach was integral to development of the new Canada - United States of America bi-national S&T cooperative programme and should serve as a sound model for further international cooperation to address public security and safety priorities.

Notes

1 For more information see www.crti.drdc-rddc.gc.ca

2 Others have championed the use of representative scenarios, or cases, to systematically and rigorously examine prevention, preparedness and response requirements as well [1].

References

[1] Danzig R. Catastrophic Bioterrorism - What is to be Done? Washington DC: Center for Technology and National Security Policy, National Defence University; August 2003.

[2] Kournikakis B, Armour, SJ, Boulet CA, Spence M, Parsons B. Risk Assessment of Anthrax Threat Letters, DRES Technical Report 2001-048. Suffield (Alberta): Defence R&D Canada; September 2001.

[3] Jernigan DB, Raghunathan PL, Bell BP, Brechner R, Bresnitz EA, Butler JC. Investigation of bioterrorism-related anthrax, United States, 2001: epidemiologic findings. Emerging Infectious Diseases 2002; Vol. 8: 1019-1028.

[4] Dull PM, Wilson KE, Kournikakis B, Whitney EAS, Boulet CA, Ho JYW. *Bacillus anthracis* aerosolization associated with a contaminated mail sorting machine. Emerging Infectious Diseases 2002; Vol. 8: 1044-1047.

[5] Dewan, PK, Fry, AM, Laserson K, Tierney BC, Quinn CP, Hayslett JA. Inhalational anthrax outbreak among postal workers, Washington, D.C., 2002. Emerging Infectious Diseases 2002; Vol. 8: 1066-1072.

[6] World Health Organization [WHO]. Public health response to biological and chemical weapons: WHO Guidance. Second Edition. Geneva: WHO; 2004.

[7] Perkins BA, Popovic T, Yeskey K. Public health in the time of terrorism. Emerging Infectious Diseases 2002; Vol. 8: 1015-1018.

Science and Technology Policies for the Anti-Terrorism Era
Andrew D. James (Ed.)
IOS Press, 2006

Engaging the Science and Technology Community in the Fight against Terrorism

Heiko BORCHERT[1]
Düsseldorf Institute for Foreign and Security Policy e.V. (DIAS)
c/o Faculty of Law, Heinrich Heine University
Universitätsstrasse 1, 40225 Düsseldorf, Germany

Abstract: The changing security environment puts a high premium on more systematic engagement of the science and technology community in order to improve relevant security capabilities. New science and technology clusters in the fields of biotechnology and life sciences are likely to be increasingly important in countering dangers stemming from the use of chemical, biological, radiological, nuclear, and explosive (CBRNE) material. Although governments will benefit from entering into long-standing partnerships with key stakeholders in these new areas, engaging them will not be easy. Specific normative, regulatory, economic and scientific disincentives need to be overcome. Besides addressing these problems, this Chapter argues that a more strategic approach to security science and technology will be necessary. To this purpose, a comprehensive map of existing capabilities and needs has to be laid out, security science and technology policies commensurate with the new risks and a nation's grand strategy need to be set up, and security capabilities and security science and technology need to be developed in tandem.

1. Introduction

The 21st century security environment deviates fundamentally from the past. Many of today's risks such as terrorism and organised crime are transnational and originate from problems within rather than between states. The rise of non-state actors ready to use force and the failure of state structures in various regions of the globe coincides with the proliferation of weapons of mass destruction and ongoing regional conflicts. The international community is required to intervene more often than in the past. Due to the power of globalisation and modern interdependencies, the consequences of these conflicts can no longer be confined to zones of crises in some distant lands. Under these conditions, the distinctions between key concepts of traditional security policy – "domestic" and "foreign", "war" and "peace" as well as "combatants" and "non-combatants" – become blurred, thus rendering them potentially dysfunctional. As a consequence, existing security institutions and the state's instruments of power need to be transformed in order to reflect these changes [1].

All of this suggests that there is an increased need to address the full spectrum of science and technology competencies that are available to aid the establishment of security capabilities commensurate with the new security challenges. This requires changes in today's science and technology policies, because new science and technology areas need to be included, and additional actors, who have so far ranked rather low on the national security agenda (such as first responders), need to be incorporated. However, science and technology is Janus-faced: it is a key enabler in achieving the necessary transformation of today's security forces and institutions, and at the same time, it is precisely the dual use character of many science and technology areas that creates opportunities for benevolent and malevolent use of the respective expertise. Given the ambivalent nature of science and technology, any discussion about its security-related contribution should focus not only on single issues such as the fight against terrorism, but should keep in mind all security risks (of man-made and natural origin). Doing otherwise runs the risk of producing stand-alone recommendations and proposals that can hardly be coordinated with activities to address other security risks.

The plea for comprehensiveness is not yet adequately mirrored in the most recent activities to step up security research in Europe. Given the pressing need to overcome civilian and military capability shortfalls in dealing with terrorism and other security risks, there is an understandable trend that favours quick wins. The Report of the Group of Personalities and the Preparatory Action of the European Commission for the European Security Research Programme (ESRP) bear ample proof of this tendency by prioritising applied technology research and development to support front-end missions of civilian and military security actors [2, 3].

This approach is not risk-free. It tends to neglect not only the contribution of basic research, but also the role of the social and cultural sciences and their contribution to understanding the political, cultural and societal aspects that underlie today's security problems. Instead, security research must be defined against the background of Europe's comprehensive understanding of security. It should also encompass science and technology to address the need to identify the relevant security risks, understand their origins, monitor their development, analyse their international and national impact, set up possible counter-strategies and define the relevant capacities and capabilities to implement these strategies.

To achieve this ambitious goal, specific disincentives need to be overcome and incentives must be strengthened in order to stimulate and engage the science and technology community. This Chapter looks at the normative, regulatory, economic and scientific environment in order to identify key enablers and obstacles. The aim is to illustrate what works and where further activities will be necessary.

2. The Overlap between Science and Technology for Military and Security Missions

Before going into detail, it should be pointed out that no consensus has yet emerged as to the degree to which military products – and the respective science and technology expertise – are suitable for the new, primarily civilian, security tasks. Whether we stress differences or similarities between the two has a strong impact not only on science and technology requirements, but also on the doctrine, organisation, training, leadership, material, and systems of the military and civilian security forces.

On the one hand, there can be no doubt that traditional defence contractors have a natural interest in entering the new homeland security markets by pulling-through their expertise and products. In some cases this makes perfect sense, as it can be assumed that, for example, information and communication technology requirements will be roughly equal. Serving military and civilian security forces with similar or the same products will help improve interoperability among them.

On the other hand, there may be areas where the use of military products and systems for civilian purposes is debatable. A case in point is protection against chemical, biological, radiological, nuclear or high-yield explosive risks (CBRNE), which, interestingly, is one of the tasks most often referred to in the discussion of a possible homeland security role for the armed forces. Here, military procedures, the higher propensity to risk taking (including death tolls), as well as different operational requirements (such as performance in case of attack versus 24 hours a day, 7 days a week performance in the civilian environment) point in the direction of adaptations or new product developments [4]. However, this does not have to be the case. The Danish Emergency Management Agency, for instance, has relied on military equipment standards in CBRN protection [5]. Following this line of argument, the Preparatory Action for the new European Security Research Programme aims at leveraging the potential of dual use technologies for applications in both domains [2]. It seems that the lack of money in general and the shrinking budgets for security tasks in particular is leading towards convergence between the domains in Europe, whereas U.S. experts tend to emphasise differences. As in the long-run this could pose a particular problem for transatlantic interoperability among military and civilian security forces, both sides must address this issue in more detail.

3. The Normative Environment

Normative incentives and disincentives make up the general framework within which the security and science and technology communities operate. A growing body of literature, from sociology to organisational studies to political science, emphasises the fundamental importance of norms, rules and principles in shaping our understanding of reality and thus also our behaviour [6, 7, 8]. Among the potential normative disincentives that could hamper cooperation, conflicting values and a lack of awareness and interaction can be singled-out as most important, because they influence all the other areas mentioned below.

Conflicting values can be found within and between public, scientific and industrial stakeholders. First of all, opaque and cumbersome public sector processes can deter private and scientific partners to develop counter-terrorism measures. Civilian industries such as biotechnology and genomics operate in a highly competitive, high-risk, fast-paced environment, where investments and decisions to pursue or discontinue must be made rapidly. By contrast, bureaucratic procedures, such as long delays while grant applications are being reviewed by funding agencies, can create uncertainty that negatively influences investment decisions and companies' economic forecasts. In addition, the public sector's overemphasis on secrecy – e.g., through classification of information, partial distribution of information and request for security clearance – can deter companies from entering into dialogue with public agencies. By contrast, a recent report by the UK House of Commons Science and Technology Committee points out that the open approach of the U.S. Government especially with regard to information sharing and declassification of information has had a positive effect of getting companies interested in developing countermeasures [9]. Second, a lack of awareness is linked to conflicting values, and it results, among other things, from the fact that new science and technology communities, such as bioscience, and the national security community do not yet interact as closely as is the case in the traditional defence-industrial complex [10, 11]. The lack of regular interaction creates personal distance and makes it difficult for actors to understand each others' interests, needs and fears. This is a particular problem in the development phases, where scientific and technological expertise would be needed to inform policy makers about how their envisaged policy actions could be backed with available knowledge and about what should be done to guide the creation of scientific and technological know-how and products [12]. There is no easy way to remedy these problems, because they are inherent in the self-understanding of the respective actors with their distinct organisational cultures.

Important normative issues also arise when we consider the moral and ethical responsibilities of scientists. Following a new line of constructivist literature in international relations theory it can be argued that norms of stigmatisation can serve as powerful tools to steer individual behaviour in directions deemed socially acceptable [6, 7,

8]. Transferring these insights to biodefence research would require the scientific community to "build a universal consensus, particularly among scientists, that the development, production or dissemination of biological weapons (...) would be regarded (...) as one of the most serious of all crimes" [13]. Against this background, self-regulation by critical industries and scientists could work as a powerful correction mechanism to avoid abuse of biodefence expertise and products. Furthermore, self-regulation could help preserve freedom of action, because "if the scientific community does not take stronger action to regulate itself, then it risks having ill-judged restrictions placed on it by politicians" [9]. Finally, governments and international organisations could build on this consensus to develop norms, rules and principles for acceptable and intolerable behaviour. This in turn could make it easier to go after rule-breakers.

In addition, normative incentives in favour of economic action to improve security should be established. Here we should distinguish between actions to increase investments in favour of corporate security and investments in research and development to meet the new security needs of governments and emergency responders. As will be argued below, the latter case requires governments to change existing and introduce new regulatory and economic incentives. In the former case, much depends on corporate actors' readiness to change their behaviour. Already now, only shortly after September 11, reluctance on the part of companies to invest in security is growing [14]. For promoting understanding of the business value that security brings to the company, it is essential for corporate security directors to have access to chief executives [15]. In addition, the financial and auditing community should start to interpret investments in corporate security as good corporate behaviour that will increase corporate resilience and thereby strengthen homeland security. If actions towards increasing corporate security are viewed as "must haves" by the financial and auditing community, managers could be influenced to adapt their calculations.

Given the dynamism and complexity of technological progress, there can be no doubt that the public sector is dependent on scientific and industrial expertise and know-how. This urgent need can be interpreted as an additional driver to overcome normative disincentives. However, as argued above, cultural differences and mutual reservations can undermine readiness to cooperate. Therefore, it might be useful to create a new playing field for collaboration. In the United States, James Petro and colleagues correctly demand the establishment of a "federally funded venue for experimentally validating biotechnology threat assessments" [16]. New forms of interaction between public, private, and scientific sectors are needed to expedite the development of new technologies for homeland security and biodefence needs. To this purpose, a new collaborative environment should aim at integrating private and scientific actors more strongly into public sector processes and thus facilitate the joint development of solutions for biodefence and counter-terrorism problems. Given the many questions surrounding the dual-use nature of bioscience and other science and technology areas, this kind of close interaction would seem ideally suited to establish a dense network of contacts among various stakeholders and to increase their involvement in policy processes and in the delivery of security-relevant capabilities. In addition, the new homeland security and anti-terrorism tasks require the science and technology community to engage the end user at all stages from research and development, to testing and evaluation and implementation [12]. A joint collaborative environment can help achieve this goal.

4. The Regulatory Environment

Regulatory incentives and disincentives set particular guidelines for science and technology activities and thus directly influence decisions to undertake or refrain from certain activities. There is thus a close connection between regulatory incentives and the economic incentives that will be discussed in the next section. Although a comprehensive discussion of the different national and international regulatory regimes that guide science and technology activities is beyond the scope of this Chapter, the lack of and the stickiness of regulations will be discussed in more detail.

First, the lack of regulations has been identified as a key reason for the absence of industry action in various areas. Of special importance are the protection of individual property rights and the lack of litigation rules for products that might be used in case of urgency but have not been tested in advance [4, 9, 17, 18]. Although it is understandable that the industry demands clear-cut regulations, the demand for government action should not be interpreted as a one-way street. It is precisely the *statu nascendi* of homeland security-relevant regulations that provides opportunities for industry to come up with proposals on how to deal with the risks of certain science and technology activities and to shape the rules that will guide corporate behaviour and frame markets in the future.

Second, existing regulations can prove too "sticky" and too cumbersome to deal with the new security risks. Lengthy approval and grant permission processes represent serious barriers for the biotechnology industry, but they also affect the government. Given the complexities of the approval process, only one in five drugs that are submitted for approval by the Food and Drug Administration (FDA) will receive final acceptance.[2] Therefore, critics contend that the government might end up buying countermeasures that could eventually fail to be approved [18, 19].

By addressing these regulatory issues and the legal risks that accompany biotechnology products, governments can provide strong incentives for their development and production [18]. The following examples illustrate this. With the signing into law of Project BioShield in 2004 the U.S. Government addressed several regulatory issues [19, 20]. Directing bioterrorism-related research becomes more flexible, because the act expedites procurement processes. Under specific circumstances, the Department of Health and Human Services (HSS) can authorise the emergency use of countermeasures that have not yet been approved by the FDA or HHS. The government will procure biomedical countermeasures worth $5.6 billion over ten years. Guaranteeing a federally funded market, which is the cornerstone of Project BioShield, was deemed necessary because of the lack of adequate countermeasures and the lack of a quantifiable market. Project BioShield provides producers with more clarity about the products that are expected to be produced, and the government guarantees that it will buy them for the Strategic National Stockpile (SNS). The SNS contains medicines and medical supplies to protect the American public. In case of need, medicines can be delivered to any state in the United States within 12 hours. So far, three contracts worth close to $1 billion have been signed under Project BioShield for the production of a new anthrax vaccine using purified recombinant protective antigen (rPA), pediatric doses of liquid potassium iodide (KI), and Anthrax Vaccine Adsorbed (AVA), a licensed anthrax vaccine.[3]

But there is also a critique. Some argue that BioShield provides too few incentives for early-stage vaccine development. This holds especially true for dual use products. So far, BioShield is explicitly designed for products with defence purposes, thus potentially hampering the marketing of products for commercial markets as well. Contrary to the intention of the act, precise information of what exactly the government will buy seems to be missing. Absent access to the government's priority list, however, the industry has a hard time estimating market prospects. Others contend that financing does not take long enough. Given drug development spans of ten and more years, the five years from FDA approval covered by BioShield are not enough. Furthermore industry sources complain that in enacting the new regulations federal offices have not made full use of the new powers for instance to accelerate the process of drug acquisition. Still others believe that vaccine stockpiling is not enough and should be complemented with post-exposure therapeutic treatment. Finally, it is said that the focus on countermeasures for known agents which are covered by BioShield will not suffice. These critics argue that the programme evokes a false sense of security and – at worst – could be misperceived by others as potentially offensive rather than defensive [21, 22, 23, 24, 25].

Specific liability and intellectual property issues were addressed in other U.S. acts. The US Homeland Security Act (Section 304)[4] stipulates that when the government orders immunisation with a government-sponsored vaccine, it will assume liability for any claims. It thus effectively takes over litigation risk from industry. One of the problems of protecting intellectual property is the erosion of the term of patents during the application period, because the term of the patent starts from the application date. The Biological, Chemical, and Radiological Weapons Countermeasures Research Act also deals with this issue[5]. The act aims at increasing incentives for private sector research to develop countermeasures to prevent and treat illnesses associated with CBR weapons attacks. According to Section 1823, the act protects biotechnology and pharmaceutical companies against erosion of the term of patents due to delays caused by government regulatory review by granting the full term of the patent.

5. The Economic Environment

The discussion of regulatory issues is closely linked with economic incentives and disincentives as particular enablers or obstacles. Although market-related incentives are welcome in many cases, they can also pose particular problems [26].

First, key characteristics of the new market for homeland security and anti-terrorism solutions, products and services can have a deterring effect. These markets are highly fragmented, and cumbersome bureaucratic processes and bidding requirements tend to favour companies with established contacts, sustained sales power and abundant financial resources. Operating in this environment not only requires patience, but also specific skills and manpower to market to public clients. Unlike the defence sector, where there is – more or less – only one key client, the new homeland security market requires companies to act at different federal and sub-national levels and to understand the requirements of different clients, such as police and law enforcement, fire fighters, emergency medical services and others. Targeting public and corporate markets at the same time is a particular challenge for small start-up companies [27]. Nevertheless, there is a growing trend among venture capitalists to push these companies to sell to the government and to adapt their strategies and sales forces accordingly [28].

Second, readiness or reluctance to enter new markets depends on growth perspectives and profit expectations. As of today, the fragmentation of homeland security products and services makes it difficult to quantify the market potential. Given budget increases, especially in the United States, the estimates assume a high growth market. However, the portion of the budget available for purchases of services, goods, equipment and research and development contracts may be much lower. The O'Gara Company, a homeland security investment firm, estimated that roughly a fifth of the 2004 homeland security budget or $7.2 billion is available for these spending categories, while the rest is eaten up by personnel and other costs [29].

These prospects must be interpreted in the light of diverging profit expectations. In 2001/02, profit before tax of successful biomedical products in the U.S. was three times that of operating margins for traditional defence contractors. In 2002, biotechnology and pharmaceutical companies that entered new markets expected annual operating margins of 28 % and 26 %, respectively, which is two to three times higher than the 7-14 % operating margins reported by traditional defence companies such as General Dynamics, Boeing and Lockheed [11]. In addition, expectations also correlate with the flow of public money. At least in Europe, it is fair to assume that public funds will remain restricted for years to come. Monies to be invested in one sector, say bioresearch, will have to be diverted from other sources, thus leading to crowding-out effects among different scientific areas. Finally, it has been speculated that the rise of money for biodefence research will increase the number of people working in the business and could therefore aggravate already existing security problems, such as the need to avoid the "brain drain" of scientists or the proliferation of sensitive research findings [30].

When thinking about economic incentives, it is important to distinguish among market forces, company strategies and government action.

First, the current way of executing new homeland security tasks and the nature of the new challenges play into the hands of steady market growth. So far, solutions for the new tasks are labour-intensive. US estimates assume that about 90 % of the $3-5 cost of screening a single passenger without checked luggage at airports amount to labour [31]. In addition, the protection of vast borders and coastlines and the improvement of cyber-security are simply too difficult and complex; improved technologies are required. Both of these trends will favour technologically advanced solutions that reduce human errors, ease the new difficulties of travel – which is a key industry in most countries around the world – and provide surveillance. By doing so, current market conditions serve as stimuli for private sector engagement in the corresponding market segments. Second, companies' market positioning and strategies will determine if and how they succeed in the new market. In order to assess business prospects, the O'Gara Company has developed the matrix shown in Figure 1 [29]. The chart distinguishes between existing capabilities to address the homeland security market (such as people, products, technologies) and the willingness of the company to leverage or improve capabilities for the new market. General interest companies in the lower left quadrant of the figure are bystanders in the new market. They can either develop their capabilities or sell them to other companies, but it will be difficult for them to serve the government directly. At the lower right, companies have capabilities for the new market that they have not yet aggressively sold. They can cherry-pick opportunities, but in the long-run they will come under pressure from more aggressive marketers. Companies in the top left quadrant are aggressively approaching the new market. They can either develop their capabilities and target subcontractors or partner with other firms to take on systems integrators. The top right corner consists of the market-makers. They set market trends, because they can offer comprehensive solutions ("system of systems") to customers. In order to hold their position, these companies need to defend their existing market share and grow at the same time to shape the market.

Third, when thinking about government action to provide economic incentives, it is important to keep in mind such aspects as the general economic situation of a country and the short- and long-term effects of government intervention [32, 33, 34]. That said, there are different options for providing economic incentives.

One option is to create a market. As discussed above, the U.S. government selected this approach with the launch of the BioShield program. By providing a stable market outlook, the government aims at stimulating companies to pass the threshold of launching R&D activities and manufacturing relevant products. The UK House of Commons Science and Technology Committee argues that it is "unrealistic to expect [biotechnology] industry to respond in the absence of a definable market" and emphasises that other industries are looking to the government for advice as well [9].

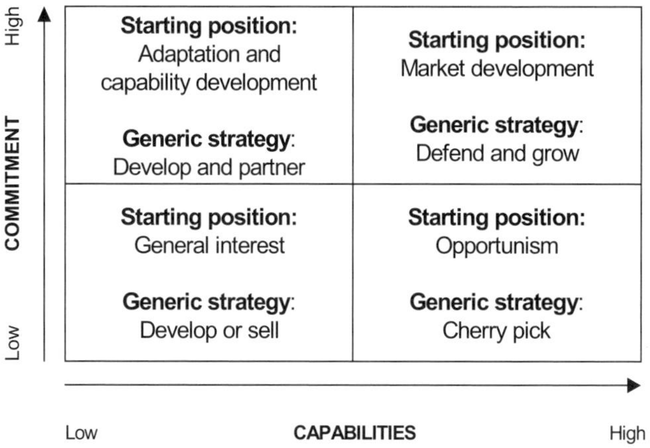

Figure 1: Positioning and strategies in the homeland security market [29].

Equally, the government can stimulate and facilitate R&D in order to support the development of security-related products in the early stages. Funds, in the form of seed money for instance, can be used to finance venture projects and support collaborative R&D. A very interesting example is the Centre for Commercialization of Advanced Technology (CCAT) based in San Diego, California. Set up as a public-private partnership, the CCAT is supported by Congress, funded by the Department of Defense and includes academic and industry partners. CCAT supports new technologies that address solutions in defence, homeland security and crisis/consequence management through the development process. Similarly, the Department of Defense activities within the Congress-sponsored Small Business Innovation Research (SBIR) Program aim, among other things, to increase private sector commercialisation of federal R&D [26]. Furthermore, increasing manufacturing capacities can be facilitated by financing the construction of manufacturing facilities and special laboratories, such as bio safety level 3 or 4 laboratories [18].

A different option is foreseen in the U.S. Biological, Chemical, and Radiological Weapons Countermeasures Research Act. According to Section 1821, a special class of stock to fund research would not be subject to any capital gains tax, and special tax credits help fund research [18]. A similar idea is also under discussion in a new European Commission proposal for science and technology. With the help of a "research tax credit" the Commission believes that private investment in research in general could be boosted, which would contribute towards achieving the research investment spending target of 3 % of GDP by 2010 [33]. A further alternative is followed by the Defense Advanced Research Projects Agency (DARPA) of the US Department of Defense. Here, the U.S. government supports high-risk projects through bearing a higher portion of the product development risk and translates support in the early stages into lower procurement costs in case of success [18].

Another way that government can act is by modifying the rules for funding [26]. In the defence sector, the U.S. government typically funds all development costs, and companies can get reimbursed for private R&D costs. In Europe, companies are expected to cover at least portions of the development costs on their own. This also holds true for the new Preparatory Action for Security Research, which foresees 75 % Community funding for industrial research and 50 % for pre-competitive development activities [36]. Given the high development risks in the fields of biotechnology or life sciences and taking into account past experiences, it seems fair to suggest that the Community and the European governments should reconsider their policy on this issue.

6. The Scientific Environment

Finally, there can be disincentives in the immediate scientific environment that can deter the academic research community from entering into new security-related activities. One of the key concerns of university-based scientists is censorship in the form of national security classification of research. Given the central role of publication in academic science, measures that threaten to limit publication prospects are likely to the subject of criticism and discourage participation in biodefence work. In order to provide guidance on this important issue, the editors of sixteen leading scientific journals adopted a joint "Statement on Scientific Publication and Security" in February 2003. While acknowledging the important contributions that scientific research can make towards tackling the new security risks, the statement also recognised that on occasions an editor may conclude that the potential harm of publication outweighs the potential societal benefits. Under such circumstances, the paper should be modified, or not be published [37]. Although the issue of censorship and security classification needs to be dealt with, the risk should also not be overestimated. According to the American Association for the Advancement of Science, only two manuscripts out of 14,000 that were reviewed in 2002 were requested to be modified for security reasons [30].

Furthermore, Smith and colleagues refer to academic scientists' dedication to pursuing fundamental basic science discoveries as a potential obstacle to addressing specific biodefence research needs. As they say:

> "Academic bioscientists are motivated by many different ambitions, but most share the goal of making fundamental contributions to the store of human knowledge… On the contrary… many academic scientists believe that biodefence research will have as its objectives applied research projects aimed at near-term development of a specific product…" [11].

These concerns underline the fact that attracting talented university scientists will require a government commitment to support biodefence research over the long term. Strategic leadership and guidance as well as continuous funding are necessary to help researchers define appropriate countermeasures to deal with biological warfare agents, to detect and defeat known and unknown pathogens, and to invent new strategies for protecting military personnel, emergency responders and civilian populations from agents of unknown properties and origin [16, 38]. In addition, more must be done in order to educate scientists about the nature of the dual use dilemma of biotechnology and other new science and technology areas [37]. This task should be managed in close cooperation with leading scientific associations and national academies. Furthermore, education programmes should also include an outreach component – a kind of "science and technology diplomacy" [39] – to enter into dialogue with scientists from countries that are presumed to be possible states of concern.

Where researchers engaged in different security-related areas have not yet joined forces to set up scientific research networks, governments should provide supporting activities, such as the Canadian CBRN Research and Technology Initiative [40]. In that case, fifteen departments and agencies have joined forces to play an active role in leading and redirecting Canada's science and technology community to address the new security risks. A strategic approach was adopted to derive science and technology targets from an overall risk assessment. In addition, laboratory clusters were formed to address capacity requirements for response, and a new project model, which is open to stakeholders from government, academia and industry, fosters collaboration. In fostering networks, special attention should be given to transnational cooperation, because it helps increase the quality of research and improve the dissemination of knowledge and results [35].

Finally, the scientific community and governments should work together to increase mutual exchange of experts by creating new opportunities for scientific experts to enter into government positions and vice versa [11]. This could stimulate the transfer of knowledge and familiarise scientific experts and public employees with each other's working environment.

7. Outlook

This chapter has argued that science and technology can make a valuable contribution to the strengthening of security capabilities. Policy makers will have to find ways of overcoming specific disincentives to such participation if the science and technology community is to be fully engaged in the fight against terrorism. The disincentives and incentives are summarised in Figure 2.

In order to fully leverage the potential, the relevant actions need to be integrated into an overall framework for the successful management of science and technology in the 21st century. Therefore, the concluding section of this Chapter looks at what should be done to approach science and technology more strategically.

7.1 Map Existing Capabilities

With few exceptions governments hardly have a comprehensive overview of existing national science and technology capabilities to address the new security risks in general and new demands in biodefence and counterterrorism in particular [38]. Informed science and technology policy depends on a broad knowledge of what is and will be available and whether and how the relevant scientific and industrial stakeholders are already in touch with each other. In order to compile the relevant information at least two things are important. First, governments must launch joint surveys of their ministries' science and technology needs with regard to security-related activities. This will help define common areas of interest, which can serve as a basis for the identification of opportunities for pooled efforts, especially in the fields of defence, homeland security and crisis/consequence management. Second, a joint working group consisting of public sector, industrial and scientific stakeholders should be established to assess the capabilities of the national science and technology sectors. Here it will be particularly important not to restrict the analysis to technology-related areas, but to include the social sciences as well. Assessment results can be compared with the needs, and resulting gaps can serve as initial building-blocks for an overall security science and technology policy [41].

7.2 Develop Security Science and Technology Policies

In most European countries there is only a weak link between existing grand strategies that outline national ambitions in foreign and security policy, the missions that need to be accomplished and the science and technology capabilities required to do so. In order to overcome this shortfall, existing processes to define national security and science and technology policies should be partially merged at the level of risk and needs assessment in order to set up national security science and technology policies [12, 40]. Without doing so, the European Security Research Programme (ESRP) will lack national counterparts, and it will be more difficult to tap into the existing science and technology reservoir. This must be done in a joint approach to include various ministries (e.g. defence, interior, foreign affairs, education and science and technology, infrastructure, economics) and relevant scientific and industrial stakeholders. Close interaction at this level is important to translate political ambitions into adequate science and technology capabilities.

	Disincentives	Incentives
Normative environment Addresses the general framework within which security and science and technology communities operate	• Conflicting values (secrecy vs. openness) can hamper cooperation • A lacking sense of urgency among key stakeholders makes it difficult to launch actions	• Strong incentives for self-regulation to avoid ill-judged regulations • Establish a new playing field for public-private interaction
Regulatory environment Sets specific guidelines for science and technology activities and thus influences decision-making	• Lack of regulations (e.g., protection of intellectual property) as a reason for absence of industry activities • Existing regulations can be too cumbersome for dealing with the new security risks (e.g., approval processes)	• Make bioterrorism-related research more flexible through expediting procurement processes (e.g., Project BioShield) • Address product liability issues (e.g., US Homeland Security Act) • Cover risk of erosion of patents during application period (e.g., US BCR Weapons Countermeasures Research Act)
Economic environment Addresses the business logic for science and technology communities to enter into new markets	• Deterring market characteristics (e.g., highly fragmented, limited market potential) • Diverging profit expectations • Increase of funds could aggravate security problems	• Create a market (e.g., Project BioShield) • Stimulate research and development and increase manufacturing capacities (e.g., DARPA, US BCR Weapons Countermeasures Research Act) • Change funding rules
Scientific environment Addresses the scientific logic for science and technology communities to enter into new areas	• Fear of censorship due to national security concerns • Interest in basic research could deter scientists from engaging in national biodefence research, development and production	• Do not overestimate risk of censorship • Governments must create long-term conditions (e.g., strategic guidance, steady flow of funds) • Create networks of research institutes to create a "critical mass" and facilitate cooperation and dialogue • Facilitate exchange of experts between government and the scientific community

Figure 2: Disincentives and incentives for engaging the science and technology community

This is needed to develop and manage science and technology capacities in line with ongoing discussions on the pooling of resources at the European level and possible role-specialisation in certain issue areas. Furthermore, governments should think about the creation of an "Office for Security Science and Technology", which could integrate stakeholders from various organisations. Most importantly, it would provide strategic guidance to implement the security science and technology strategy, thereby helping to coordinate procurement programmes, science and technology programmes and other national economic actions undertaken to increase competitiveness. The office would provide a single-point of contact for requests from the science and technology community, could organise information events to build awareness for important issues and help advance the security science and technology agenda in the public domain.

7.3 Harmonise Capability Development and Security Science and Technology

A similar approach will also be required at the European level. Different initiatives such as the European Capabilities Action Plan and ESRP have been launched and new institutions such as the European Defence Agency (EDA) established in order to address existing capability shortfalls and to overcome the artificial separation between civilian, security and defence science and technology. Comparable to national endeavours, a European security white paper will be needed to translate the assumptions of the European Security Strategy into more operational plans for different security instruments available via the Common Foreign and Security Policy and the European Security and Defence Policy. Security science and technology policy must then be established as a horizontal process that cuts across these areas and includes European (e.g. Commission, Military Staff, Solana Office, Council, EDA) and national stakeholders. Of key importance will be the future relationship between the new EDA and the Commission. While the EDA concentrates primarily on defence aspects, the scope of the Commission is broader. This is a problem with regard to implementing the EU's comprehensive security agenda. Thought should thus be given to enlarging the thematic thrust of the EDA. The EDA should also address the needs of emergency responders. Given the need for close coordination between civilian security forces and military forces in case of potential terrorist attack involving CBRNE risks, interoperability becomes a key requirement. Furthermore, military and civilian forces increasingly draw on similar technologies to accomplish their missions. In order to address these similarities, it seems wise to analyse lessons that can be learned from the U.S. government-sponsored CCAT mentioned above and from DARPA as well as the sister organisation in the Department of Homeland Security. Equally, the personal composition of the EDA steering board should be modified as well. The current focus on Ministers of Defence or their representatives and a representative of the Commission is too narrow. Non-defence science and technology stakeholders must also be included in order to facilitate the mutual exchange of expertise between defence and non-defence science and technology.

Notes

1 I thank Ellen Russon for editorial assistance

2 Food and Drug Administration. From Test Tube to Patient: Improving Health Through Human Drugs, Washington D.C., September 1999; cited in [19].

3 See: <http://www.hhs.gov/ophep/bioshield/PBPrcrtPrjct.htm> (accessed 7 June 2005).

4 See: <http://frwebgate.access.gpo.gov/cgi-bin/getdoc.cgi?dbname=107_cong_bills&docid=f:h5005enr.txt.pdf> (accessed 27 November 2004).

5 See: < http://frwebgate.access.gpo.gov/cgi-bin/getdoc.cgi?dbname=108_cong_bills&docid=f:s666is.txt.pdf> (accessed 27 November 2004).

References

[1] Borchert H. Vernetzte Sicherheitspolitik und die Transformation des Sicherheitssektors: Weshalb neue Sicherheitsrisiken ein verändertes Sicherheitsmanagement erfordern. In: Borchert H, editor. Vernetzte Sicherheit. Leitidee der Sicherheitspolitik im 21. Jahrhundert. Hamburg: Verlag E.S. Mittler & Sohn; 2004: p 53-79.
[2] Commission of the European Communities [EC]. Research for a Secure Europe. Report of the Group of Personalities in the Field of Security Research. Luxembourg: Office for Official Publications of the European Communities; 2004.
[3] Commission of the European Communities [EC]. On the Implementation of the Preparatory Action on the Enhancement of the European Industrial Potential in the Field of Security Research: Towards a Programme to Advance European Security through Research and Technology, COM(2004) 72 final, Brussels: EC; 3 February 2004.
[4] Albright PC, Dockery H. A framework for homeland security research and development: the United States perspective. In this Volume; 2006.
[5] Daldgaard-Nielsen A. Homeland security and the role of the armed forces: a Scandinavian perspective. In: Borchert H, editor. Weniger Souveränität – Mehr Sicherheit. Schutz der Heimat im Informationszeitalter und die Rolle der Streitkräfte. Hamburg: Verlag E.S. Mittler & Sohn; 2004: p 59-75.
[6] Wendt A. Social Theory of International Politics. Cambridge: Cambridge University Press; 1999.
[7] Adler E. Constructivism in International Relations. In: Carlsnaes W, Risse T, Simmons BA, editors. Handbook of International Relations. London: Sage; 2002. p 95-118.
[8] Ruggie R. Constructing the World Polity. Essays on International Institutionalization. London: Routledge; 1998.
[9] House of Commons Science and Technology Committee (UK). The Scientific Response to Terrorism. Eighth Report of Session 2002-03. London: The Stationery Office; 2003.
[10] Kwik G, Fitzgerald J, Inglesby TV, O'Toole T. Biosecurity: responsible stewardship of bioscience in an age of catastrophic terrorism. Biosecurity and Bioterrorism 2003; 1: 27-35.
[11] Smith BT, Inglesby TV, O'Toole T. Biodefense R&D: anticipating future threats, establishing a strategic environment. Biosecurity and Bioterrorism 2003; 1:193-202.
[12] Dockery H, Albright PC. Creating a paradigm for effective international cooperation in homeland security technology development. In this volume; 2006.
[13] O'Toole T. Institutional issues in biodefense. In: Roberts A, editor. Governance and Public Security. Syracuse: Campbell Public Affairs Institute; 2002. p 99-113.
[14] Flynn ST. The neglected home front. Foreign Affairs 2004; 83: 20-33.
[15] Cavanagh TE. Security in Mid-Market Companies: The View from the Top, Executive Action No. 102. New York: The Conference Board; 2004.
[16] Petro J, Plasse TR, McNulty JA. Biotechnology: impact on biological warfare and biodefense. Biosecurity and Bioterrorism 2003; 1: 161-168.
[17] Bush K. Improving and creating opportunities to develop vaccines against bioterrorism threats – an industry perspective. Testimony before the Senate Committee on Health, Education, Labour, and Pensions, Health Subcommittee, Washington, DC, 30 January 2003. Available at: http://health.senate.gov/testimony/007_tes.html (accessed 27 November 2004).
[18] Gottron F. Project Bioshield. CRS Report RS21507. Washington, DC: Congressional Research Service. April 28, 2003 (Updated December 27, 2004).
[19] Dominguez A. BioShield shortcomings subject of industry conference. USA Today 9 June 2004. Available at: www.usatoday.com/tech/news/techpolicy/2004-06-09-bioshield-progress_x.htm (accessed 27 November 2004).
[20] Project Bioshield. Fact Sheet by the United States Department of Health and Human Services, 21 July 2004. Available at: http://www.hhs.gov/news/press/2004pres/20040721b.html (accessed 27 November 2004).
[21] DeFrancesco L. Throwing money at biodefense. Nature Biotechnology 2004; 4: 375-378.
[22] Qualters S. BioShield: The $5.6 billion few companies vie for. Boston Business Journal, 15 November 2004. Available at: http://boston.bizjournals.com/boston/stories/2004/11/15/story5.html (accessed 17 June 2005).
[23] Dando M. The United States National Institute of Allergy and Infectious Diseases (NIAID) Research Programme on Biodefense. A Summary and Review of Varying Assessments. Bradford: Department of Peace Studies; 2004.
[24] Madigan S. Biodefense Industry Grumbling Over HHS Handling of Germwar Priorities. CQ Homeland Security 23 May 2005.

[25] Davis JH. Testimony before the United States House of Representatives Committee on Government Reforms, Subcommittee on National Security, Emerging Threats, and International Relations, 14 June 2005. Available at http://reform.house.gov/UploadedFiles/Dr.%20Davis%20Testimony.pdf (accessed 21 June 2005).
[26] James AD. U.S. Defence R&D Spending: An Analysis of the Impacts. Rapporteur's report for the EURAB Working Group ERA Scope and Vision. Manchester: PREST; 2004.
[27] Keinan T. Investing in homeland security technologies: a venture capital perspective. Tel Aviv: Giza Venture Capital; 2003. Available at: http://www.altassets.com/cgi_local/MasterPFP.cgi?doc=http://www.altassets.com/casefor/sectors/2003/nz3016.php (accessed 25 November 2004).
[28] Vaida B. Venture capitalists urge small tech firms to enter government market. Covexec.com 14 April 2003. Available at: http://www.govexec.com/dailyfed/0403/041403td1.htm (accessed 26 November 2004).
[29] Gould S, Beckner CH. The Homeland Security Market. Corporate and Investment Strategies for the Domestic War against Terrorism. Washington, DC: The O'Gara Company; 2003.
[30] Krönke M. Die Dialektik der Wissenschaftsfreiheit vor dem Hintergrund der Bioterrorismus-Bekämpfung. Bundesgesundheitsblatt Gesundheitsforschung und Gesundheitsschutz 2003:46;1010-1013.
[31] Homeland Security Research. Homeland Security Industry – Trends. Washington, DC: Homeland Security Research; 2003. Available at: http://www.homelanddefensestocks.com/Companies/HomelandDefense/News/HSRC-HomelandSecurity-TheBroadPicture10-2003.pdf (accessed 26 November 2004).
[32] Andersson JJ, Malm A. Distribution of responsibilities and money in dealing with societal security, public safety and emergency management. Paper for the 6th International CRN Expert Workshop; Stockholm; 24 April 2004.
[33] Orszag PR. Homeland security and the private sector. Testimony before the National Commission on Terrorist Attacks Upon the United States, Washington DC; 19 November 2003.
[34] Brück T. Assessing the economic trade-offs of the security economy. In: The Security Economy. Paris: OECD; 2004: p101-126.
[35] Commission of the European Communities [EC]. Science and Technology, the Key to Europe's Future – Guidelines for Future European Union Policy to Support Research, COM(2004) 353 final, Brussels; 16 June 2004.
[36] De Smet P. Programme of Work PASR-2005. Presentation for the Meeting on the Programme of Work for 2005 of the Preparatory Action in the Field of Security Research, Brussels; 8 November 2004.
[37] National Research Council (US) [NRC]. Biotechnology Research in an Age of Terrorism: Confronting the Dual Use Dilemma. Washington, DC: NRC; 2003.
[38] Carlson R. The pace and proliferation of biological technologies. Biosecurity and Bioterrorism 2003; 1: 203-214.
[39] United Nations Conference on Trade and Development. Science and Technology Diplomacy. Concepts and Elements of a Work Program. New York: United Nations; 2003.
[40] Boulet CA. Development of a science and technology response for CBRN terrorism: The Canadian CBRN Research and Technology Initiative. In this volume; 2006.
[41] Pankratz T, Vogel A. Der Aufbau sicherheitspolitischer Fähigkeiten und der Beitrag von Wirtschaft und Wissenschaft: Status quo der Sicherheitsforschung in Österreich. In: Borchert H,editor. Potentiale statt Arsenale. Sicherheitspolitische Vernetzung und die Rolle von Wirtschaft, Wissenschaft und Technologie. Hamburg: Verlag E.S. Mittler & Sohn; 2004. p 95-111.

Part 4

International Cooperation

Science and Technology Policies for the Anti-Terrorism Era
Andrew D. James (Ed.)
IOS Press, 2006

Creating a Paradigm for Effective International Cooperation in Homeland Security Technology Development

Dr. Holly A. DOCKERY
Special Assistant for International Policy
Science and Technology Directorate
U. S. Department of Homeland Security

Dr. Penrose C. ALBRIGHT
Assistant Secretary
Science and Technology Directorate
U. S. Department of Homeland Security

Abstract. This Chapter sets out the opportunities and challenges for effective international cooperation in counter terrorism technology development from the point of view of the U.S. Department of Homeland Security (DHS). Incentives for international cooperation on counter terrorism technologies are considered as are some of the non-technical influences that may be important in structuring such cooperation. The Chapter advocates a strategic approach to international collaboration and proposes a potential global research agenda.

1. Introduction

There are many obvious reasons to pursue international cooperation in support of counterterrorism technology development. The capacity expansion allowed by leveraging scarce monetary, people, and facility resources to address the range of threats is one major incentive. Closer collaboration also helps ensure that consistent and effective suites of countermeasures are deployed worldwide. The desired result is to provide a base level of protection that will benefit all nations. However, there are many issues – outside of the scientific realm – that must all be taken into consideration. These matters – policy considerations related to privacy and secrecy, practical issues including governance, legal concerns such as intellectual property rights, and considerations such as cost and operational environments – are constraints on any domestic science and technology programme with a homeland security focus. Such considerations become even more difficult to deal with in an international environment where widely varying entities must be represented. Indeed, it may be necessary to engage international agencies outside of the science community in helping resolve these problems.

Foremost, to become important and effective components of their country's approach to homeland security, science and technology providers must adopt a new paradigm for prioritising their efforts. In short, a more strategic perspective on scientific development is essential to ensure a focus on the highest-payoff technologies. Today, the changing requirements placed upon the technology system as federal security missions become more centralised in the civil sector requires building approaches and constituencies unlike those furnished by military models. The engagement of all stakeholders in the planning and implementation of technology programmes will facilitate the movement away from a technology push to a more capability driven process. Developing a common methodology that allows each country to determine its priorities in a more rigorous manner is also an important element of success. Developing mechanisms for bilateral and multilateral dialogue to identify and establish research, development, testing, and evaluation priorities and then to distribute work among the various partners is a long-term effort, but one that can be built iteratively. Establishing a globally accepted methodology and associated prioritisation criteria can be used as a guideline for initiating a strategic capability driven approach. In the interim, as a coordinated approach is developed, it may be useful to propose some obvious priority areas for international collaboration that would allow the dialogue to begin immediately.

2. Incentives to Collaborate on Homeland Security Science & Technology

For every nation, achieving an acceptable degree of "homeland security" is an effort that requires international cooperation on many fronts. Construction of technology-enhanced security systems is one key example. In the context of the modern world, we can no longer effectively protect our citizens by erecting a contemporary equivalent of Hadrian's Wall. Our economies and our quality of life are sustained by an uninterrupted high-volume flow of goods and people back and forth through our ports of entry. Emergencies do not recognise geopolitical boundaries and so we must be prepared to work directly with our neighbours to respond to natural and man-induced disasters and to maintain common infrastructure components. We are all increasingly reliant on an international cyber network that has no borders. Therefore, foreign partnerships are critical to decreasing the permeability of our shared borders to undesirable people or goods by increasing the effectiveness of the global technology systems that we rely on to provide layers of defence[1]. In essence, public security is better accomplished by adopting a "neighbourhood watch" approach that places many eyes on the problem around the world. By doing so, the global community can focus its efforts on developing flexible and adaptive homeland security technologies for global deployment.

2.1 Building a Balanced Defensive System

It is in the interests of the global community to establish roughly equivalent homeland security systems in particular areas of concern. As the sophistication of the systems providing public security becomes more unequal, the threat will be progressively directed to the softer targets. That said, a multilayered defence system complicates the task of the terrorist. Robust defensive systems will force adversaries along longer pathways to a target, will require them to involve more individuals in their conspiracies – including more technically competent people – will cost them more in time and funds, and will make their communications more difficult[2]. All of these obstacles, in turn, make our adversaries more susceptible to interdiction by traditional law enforcement mechanisms, as well as by the more specialised homeland security defensive systems.

Many of the systems that can be developed can also enhance other aspects of public safety and security. For example, bio-surveillance and medical countermeasures will be equally effective for natural disease outbreaks in humans and in the agricultural sector. In addition, container tracking initiatives will lead to greater efficiency in transportation of goods. Hence, the raising of the base level of every nation in terms of homeland security technology provides universal benefits to the global community.

2.2 Leveraging Limited Resources

The international research community has begun to apply its considerable and diverse expertise to countering terrorism. There are a number of strong drivers for worldwide scientific coordination, collaboration, and information sharing to support the counterterrorism mission. In particular, developing the technologies that provide effective countermeasures to weapons of mass effect is an enormous undertaking. Thus, advances in mathematics, physics, chemistry, and biology are needed to establish and confirm the assumptions underlying our technology-based countermeasures.

Yet no individual country has the capacity to adequately support the comprehensive scientific programmes needed to address the entire suite of countermeasures to high consequence means of attack or the tools required to contain the criminals that would develop and use these means. Our individual monetary resources are simply insufficient.

Equally important is the fact that there are only a limited number of people with the necessary expertise and facilities capable of supporting the specialised activities related to WMD research. Some countries have been the first hand recipients of terror for many years. Their unique experience provides invaluable information on lessons learned that should be recognised as crucial to understanding the problem and the effectiveness of specific solutions. As the threat has evolved to be driven by high-tech instruments of attack, it has become important to find ways of coordinating the limited pool of scientists and engineers with the expertise and experience to counter the threat.

However, as we look across the many programmes throughout the world, we find many countries are trying to confront the same problems. It is clearly of mutual benefit to determine how to best leverage both funds and people to diminish duplication. It will also be useful to take advantage of differences in approaches in solving these very hard problems. Building a framework that will encourage a strategic and integrated scientific approach to the shared threat will benefit the global community. Thus, a strategy for developing and implementing international cooperative activities must focus on leveraging the global science and technology enterprise to solve homeland security problems and also on ultimately providing leadership to set the long-term research, development, testing and evaluation agenda for that enterprise.

3. Non-Technical Influences on S&T Cooperation for Homeland Security

Scientific considerations are often not the dominant influence on which technology is ultimately deployed in homeland security settings. Decision makers and users have procedural, operational, and cost considerations. Cultural issues, domestic laws, existing infrastructures, and traditional alliances are among the policy pressures influencing the decision to pursue a specific technology solution. Perhaps even more influential is the perception of risk that any nation or government has concerning its relative vulnerability to various means of attack. To be effective in setting the agenda for international cooperation and collaboration, the science and technology framework will have to take into account a number of non-technical factors.

3.1 User Capability Gaps

Homeland security requires "applied" science. A primary consideration for all technology system developers is how to best understand and capture what a decision maker needs to know and on what timeframe. Decision makers range from senior leaders deciding how to best institute a nationwide aviation security system to a city mayor deciding how to evacuate a city to a customs agent determining when to unload a suspicious vehicle. Each decision takes place on a different time scale and has to take into consideration different criteria. The consequences of a mistake in judgment lead to vastly different consequences.

The advent of catastrophic terrorism has required localised civil resources to consider how to more effectively operate in a broader world. They must be prepared to connect with federal, state, and other local providers both in response to a crisis – but also as part of a network that continuously exercises protective and preventative measures. Equipment must be interoperable, cost-efficient, suitable for both the operator and for the operational environment, and sustainable. Communication must be timely and appropriately informative. Security technologies that greatly impede travel and trade may be rejected if the balance between efficiency and precaution is deemed unacceptable. These elements and others will impact technology programmes.

To be effective, the science and technology enterprise must move away from a "technology push" mode and toward a paradigm that aggressively engages the end users at all stages of the research, development, testing, and evaluation cycle. The answer is not to simply adopt user requirements. Rather, scientists must help provide guidance on how technology does – or does not – mitigate a specific capability gap. This is particularly true within the WMD countermeasures realm. Because a large percentage of our resources have been spent on improving military capabilities to operate in these types of environments, there is a strong push to simply recycle and reuse those same technologies in civil applications. However, even though there is a common underlying scientific base, the civilian applications are radically different, as are the funding streams for purchase and the infrastructure for upkeep and training. Establishing a partnership between technology providers and users will facilitate an understanding of the strengths and limitations of various solutions that will translate into an effective homeland security system.

3.2 Policy Issues

Policy and legal considerations constrain the technologies used and their implementation. However, technology developers must help inform policy decisions so that technology capabilities are optimised. Formal negotiated instruments between or among partner countries are generally necessary to allow federal entities to collaborate in any substantive way on technology development. Laws surrounding intellectual property rights, the ability to share classified information, the ability to exchange funds, staff, or test items all govern the form and content of international cooperative science and technology projects. The breadth and depth allowed for cooperation revolves on the degree of trust that nations have for each other. It is easier to institute programmes among allied countries than to reach outside of the traditional networks. Cultural concerns are reflected in the policies that are developed. An example is the reluctance of many countries to collect, much less to share, fingerprints. Other constraints arise when ownership and governance of shared technology systems are pursued.

Partnering with international organisations – such as the Organisation for Economic Cooperation and Development (OECD) and its Working Party on Information Security and Privacy (WPISP) that promotes an internationally coordinated approach to policymaking in security and protection of privacy and personal data – can help the technical community develop tools that are more broadly accepted. Conversely, technologists can help the policy makers understand the capabilities that advanced technologies can provide when tailored to their needs. Other organisations or coalitions exist and could also provide opportunities for working the complex problem of connecting policy with technology.

4. Establishing a Framework for Effective International Collaboration

There are a number of models for international cooperation. These range from fora provided by technical professional societies that focus on academic exchanges, to loose multi-lateral information sharing arrangements and from very formal bi-lateral and multi-lateral instruments allowing broad collaborations to strictly one-way contractual arrangements. In the security realm, these agreements tend to be driven by military needs. Joint science and technology programmes have focused on defence capability development – but most do not provide the most desirable framework to allow the balancing of secrecy against sharing of information and technologies.

The evolution to a civil security environment is still ongoing. Most science and technology collaboration initiatives – outside of a few programmes – are not sustained and programmatically focused efforts nor can they be demonstrated to address strategic needs.

For the United States, there exist a number of government-to-government and agency-to-agency agreements supporting bi-lateral and multi-lateral programmes. However, because the homeland security mission has led to a change in government structure and reporting mechanisms, the entire suite of desired topics is not currently covered in any individual agreement. Even if an amalgam of individual agreements could be identified to cover the scope of work that supports the science and technology goals of the Department of Homeland Security (DHS), without umbrella agreements the establishment and governance of an integrated strategic set of programmes is improbable.

5. Instituting a Strategic Approach

Prior to the formation of the DHS, the priorities for research and development activities that applied to U.S. homeland security needs were established by various agencies and investigators based on their individual interests. However, there was a lack of a coordinated approach that served to systematically focus resources on the highest priority issues or to ensure that user capability needs were adequately captured. International activities were generally considered peripheral to the primary work of an organisation so, even on high-priority tasks, progress tended to be very slow and the deliverables were often non-essential to the projects that they supported. The result is that no mechanism exists to determine if these programmes contribute to an overall increase in security in any quantifiable way. There has also been no urgency attached to accomplishing project goals. This lack of a strategic approach to research, development, testing and evaluation for homeland security applications in the past has lead to a generally unsatisfactory return on investment[3]. There is clearly a need, therefore, to assess the current and future suites of opportunities for engagement in the context of a strategic framework, and then to plan, perform, and integrate the task into mainstream activities.

The strategic framework must pursue not only the best science but also the best science for the task. It must formalise the incorporation of operational capability requirements. It must also explicitly consider the policy constraints. In order for programmes supporting international collaborations on homeland security science and technology to be effective, they must be viewed primarily as an additional mechanism for accomplishing established priority goals. The specific threats, vulnerabilities, and gaps – and the relative priorities for establishing collaborative engagements with foreign entities – should be driven by an overarching coordinated assessment of risk.

5.1 Establishing Strategic Priorities for Homeland Security Science and Technology

Each nation will have a somewhat different threat-driven suite of priorities and each will have different capabilities and resources to address their highest priority needs. Developing a globally accepted risk assessment is a massive undertaking and obtaining consensus on the priority suite of activities may seem to be a daunting task. However, as each nation develops the criteria against which to test its readiness to protect its citizens against terrorism, it will reach an understanding of where technology can play the most significant role in decreasing its important vulnerabilities. There will be a greater incentive to cooperate in the largest areas of overlap where consistency and transparency is most critical.

Codifying a general approach that every individual country can use might be the first step in the effort to develop an international strategy for homeland security science and technology. To be effective, the general approach must accomplish several important goals. It must provide a forum for cross-agency coordination. Because the responsibilities for homeland security will never be held exclusively by one agency, accomplishing this coordination is very important. It must provide an explicit method of connecting the user community with the technology developers. Without an understanding of their capability needs and operational environments, the application of tools developed will at best not be optimised – at worst the tools will be completely ineffective.

As the process matures, it must eventually incorporate standards and test protocols that ensure confidence in the system and its components and feedback loops that validate the effectiveness of the technologies once they are deployed. Finally, it must include mechanisms that revisit the system on an as-needed basis to update the system as technology streams and threats evolve.

As each country understands its relative priorities, it can begin the process of interacting with other nations and international agencies to plan and implement mutually recognised priority programmes.

5.2 A Model for Prioritised Collaboration

The Canada – U.S. Public Security Technical Program (PSTP) provides a general template for the approach to international science and technology cooperation efforts. A Bi-National Steering Committee was created in recognition of the fact that both countries had an enormous number of on-going technology development programmes that could support the homeland security mission. However, many programmes were clearly duplicative. Also, the results of the projects were generally invisible to many of the users – and other investigators. Most importantly, there was no confidence that this large expenditure of resources was actually resulting in any demonstrable decrease in risk. The Committee was tasked with developing an approach and mechanism to coordinate and manage these activities at a federal level.

The PSTP was the mechanism formed to accomplish the goals of the Bi-National Steering Committee. The participants of the PSTP are currently integrating their respective ongoing and future public safety and security science and technology collaboration into a single, overarching bi-national strategy to ensure efficient and effective use of joint resources. The goals of the PSTP activities are to contribute to the enhancement of those national public safety and security capabilities that will provide demonstrable and significant reduction of risk to both countries, and to develop and implement science and technology cooperation that will enable those capabilities. A Strategic Plan has been written and adopted to provide guidance on the balance of investment between science and technology activities to address immediate capability needs and those intended to enable future capabilities.

The PSTP's collaboration priorities derive from what will continue to be an evolving process to develop a coordinated assessment of public safety and security risks to each nation, and a parallel process to prioritise the gaps in national capabilities to address those risks through expert opinion and analysis.

The PSTP recognises the ultimate need for adopting an all-hazards approach to the development of science and technology solutions. However, as the programme begins, priority has been given to those events driven by terrorist or criminal activity. The working groups reflect the judgment that countermeasures to weapons of mass effect are where the largest gains in risk reduction must be accomplished in the near term. The working groups include: CBRNE (Chemical-Biological-Radiological-Nuclear-Explosive) prevention, detection, response and recovery; Critical Infrastructure Protection; Detection and Interdiction (primarily supporting border related law enforcement); and Systems Integration, Standards, and Analysis. The latter group is the one in which the cross-cutting issues are addressed – particularly those associated with the coordinated risk assessment and those that provide broad direction for the research, development, test and evaluation outcomes being sought in the longer term.

The PSTP's strategy includes providing the linkage between the "customers" in the user communities at the federal, state/provincial, local levels of government and in the private sectors, and to the nations' science and technology "suppliers" in government, academia, and industry. The PSTP framework has also allowed the identification of opportunities for leveraging resources. As programme plans have been shared and redundancies recognised, programme funds have been reallocated to allow a wider range of goals to be accomplished by both nations.

6. Setting the Global Research Agenda for Homeland/Public Security

Gaining a consensus view on the general areas that will most benefit from a consistent set of international programmes is an important step forward. Below are listed a few areas in which widespread international cooperation in science and technology is likely to provide high levels of risk reduction for each participating country. It is not an exhaustive list nor is it a list that should replace a strategic planning effort. However, it does represent areas that would greatly benefit from a concentrated multilateral effort.

6.1 Standards

One major area that would benefit greatly from a coordinated approach is the development of an integrated performance measurement infrastructure for homeland security countermeasures – supported by standards and test protocols to ensure the effectiveness of all of the components of the homeland security system. These components include equipment, information, analyses, personnel, and systems. Standards are essential in providing consistent and verifiable measures of technology effectiveness in terms of: basic functionality; appropriateness and adequacy for the task; interoperability; efficiency; and sustainability.

6.2 Cyber Security

Cyber security invariably arises as a major worldwide concern because of the widespread reliance on information technology (IT) in all sectors of our society and because of the ubiquity of IT and its interconnectedness. Information technology vulnerabilities are widespread characteristics of our shared infrastructure and severe consequences can occur if these vulnerabilities are exploited. An adversary may choose to conduct a cyber attack because he can attack from a distance with only modest technical sophistication and financial investment. Coordinated efforts must be directed at improving the security of the existing cyber infrastructure, and providing a foundation for a more secure infrastructure in the future.

6.3 Document and Traveller Authentication

At present, the security of international travel depends mainly on paper documents and manual review, limited intergovernmental data sharing, and uncoordinated national policies on privacy and information security[4]. The authenticity of travel documents is very difficult to verify. Even more challenging is establishing a high degree of certainty in the identity of the person using the document. This is problematic both because of concern about the global, uncontrolled movement of known as well as suspected terrorists and also because of the prevalence of lost, stolen, or fraudulent passports.

A number of parallel efforts are underway to establish a well-designed, globally coordinated approach, which relies on electronic databases, electronically based identification and validation techniques, and automated screening procedures, to enhance international travel security while enabling travel convenience and ensuring privacy. An example is the Enhanced International Travel Security (EITS) concept being pursued by the U.S. However, the design, implementation, and continued maintenance of such a system must consider not only the technical aspects of ensuring constant, immediate, and secure information sharing but also the policy and procedural concerns related to access to, and privacy of, such information. Various levels of trust in the system by the countries contributing or seeking data will ultimately affect database design – namely consolidated or distributed or a combination – and the structure of the international policies determining and managing operation of such a capability – whether through a series of bilateral agreements or a single or several multilateral agreements.

6.4 Detection of Threats Targeting Transportation

Methods for screening air, sea, and land based transport vehicles are generally ineffective and/or cost-prohibitive. Research and development is necessary to create integrated, rapid, accurate, transparent non-intrusive inspection and associated information systems for the point and stand-off detection of CBRNE materials, illegal drugs, illegal aliens, and other contraband in cargo loaded trucks, in containers, on maritime vessels, and in aircraft[5]. However, without international application of these tools, the global transportation system can be compromised.

7. Summary

There are a great many challenges to developing a unified suite of technologies to counter terrorism on a worldwide basis. In the technical arena there must be a shift in the way that scientists and engineers view their role in providing homeland security countermeasures. Basic science must continue to support our ability to understand the underlying mathematical, physical, and biological principles. However, attempts to push technology into the hands of the users without understanding how decisions are made or the operational requirements and constraints are at best likely to be ineffective – at worst they will actually decrease system effectiveness. Many examples of the latter are becoming obvious as we critically analyse current technology deployments. Thus, a challenge exists in convincing the technology providers to elicit detailed information on capability needs – not just requirements – from each potential user in order to develop effective solutions.

Policy issues – particularly those related to privacy and security – also drive technology development. The science community does not traditionally interact with the policy makers in the developmental phases. Policy then becomes reactive to technology rather than informing and guiding its creation. Organisations that can aid in overcoming policy obstacles will have to be engaged.

Even though developing a balanced and integrated security framework with consistent technology approaches is clearly of universal benefit, developing a shared vision and gaining consistency among the various nations will continue to present the biggest hurdle. Developing a unified general approach to developing priorities can guide the science and technology community in providing the homeland tools. The results of this assessment will help us jointly set the research agenda and leverage resources to provide demonstrable increases in the security of the world's population.

Notes

1 As the 9/11 Commission Report noted: "Exchanging terrorist information with other countries, consistent with privacy requirements, along with listings of lost and stolen passports, will have immediate security benefits... We should also work with other countries to ensure effective inspection regimes at all airports. Recommendation: The U.S. government cannot meet its own obligations to the American people to prevent the entry of terrorists without a major effort to collaborate with other governments. We should do more to exchange terrorist information with trusted allies, and raise U.S. and global border security standards for travel and border crossing over the medium and long term through extensive international cooperation" [1, pp. 389-390].

2 The 9/11 Commission Report observed: "Defenses cannot achieve perfect safety. They make targets harder to attack successfully, and they deter attacks by making capture more likely. Just increasing the attacker's odds of failure may make the difference between a plan attempted, or a plan discarded. The enemy also may have to develop more elaborate plans, thereby increasing the danger of exposure or defeat" [1, p.383].

3 The 9/11 Commission Report commented: "Our strategy should also include defenses. America can be attacked in many ways and has many vulnerabilities. No defenses are perfect. But risks must be calculated; hard choices must be made about allocating resources. Responsibilities for America's defense should be clearly defined. Planning does make a difference, identifying where a little money might have a large effect" [1, p. 364].

4 As the 9/11 Commission Report noted: "Internationally and in the United States, constraining terrorist travel should become a vital part of counterterrorism strategy. Better technology and training to detect terrorist travel documents are the most important immediate steps to reduce America's vulnerability to clandestine entry. Every stage of our border and immigration system should have as a part of its operations the detection of terrorist indicators on travel documents. Information systems able to authenticate travel documents and detect potential terrorist indicators should be used at consulates, at primary border inspection lines, in immigration services offices, and in intelligence and enforcement units" [1, p. 385].

5 In the words of the 9/11 Commission Report: "The most powerful investments may be for improvements in technologies with applications across the transportation modes, such as scanning technologies designed to screen containers that can be transported by plane, ship, truck, or rail. Though such technologies are becoming available now, widespread deployment is still years away" [1, p. 392].

References

[1] The 9-11 Commission Report, Final Report of the National Commission on the Terrorist Attacks Upon the United States, Washington DC: U.S. Government Printing Office; 2004.

Science and Technology Policies for the Anti-Terrorism Era
Andrew D. James (Ed.)
IOS Press, 2006

Enhancing Transatlantic Cooperation on S&T for Homeland Defence and Counter-Terrorism

Richard A. BITZINGER[1]
Asia-Pacific Center for Security Studies
2058 Maluhia Road
Honolulu, HI 96815-1949, U.S.A.

Abstract. Terrorism is increasingly a transnational threat, requiring transnational responses. Consequently, there are many potential incentives for expanding international cooperation in critical areas of science and technology (S&T) as it applies to developing, producing, and deploying systems for homeland defence and counter-terrorism. Europe has made considerable progress in promoting pan-European cooperation in the area of defence-related S&T, and these experiences and efforts could be useful in expanding transnational cooperation in the area of developing new technologies for combating terrorism. Given the United States' longstanding proclivity towards protectionism in its defence industrial base, it will be more difficult to expand transatlantic collaboration in the area of S&T for homeland defense and counter-terrorism. This Chapter concludes with some recommendations for overcoming these impediments, including reforming export controls, reducing controls over security of supply and information, and encouraging new institutional collaborative S&T initiatives on both sides of the Atlantic Ocean.

1. Introduction

It is no great revelation to state that terrorism is increasingly a transnational threat, and perhaps the greatest transnational security concern of the early 21st Century. Many leading terrorist groups – such as Al-Qaida or Jemaah Islamiya – are global (or at least regional) in their organisation, scope, and operations. Transnational terrorist organisations have struck recently in the United States, Spain, Africa, Bali, and, of course, the Middle East. Consequently, it stands to reason that our responses to terror should also be international. As one Western analyst has put it:

> "Modern terrorism isn't constrained by national borders; therefore the counter-terrorist response cannot afford to be either. It requires regional and international cooperation through organizations such as the UN, G8, ASEAN and APEC among others. Such cooperation is necessary to ensure the effective outlawing of terrorism organizations and associated groups, stronger restrictions on the financing of terrorism and greater assistance in nation-building in failed states where terrorism might breed and consolidate its support. It also involves providing assistance in capacity-building measures to countries that lack the resources or even the will to effectively combat terrorism" [1, p. 5].

In addition, one could add that, among the allies in the global war on terror, there should be increased cooperation in several areas of homeland defence and counter-terrorism, particularly in the gathering and sharing of information on terrorist groups and activities, and in engaging in counter-terrorist law enforcement and military operations.

Further upstream, those allies may also choose to cooperate more closely and more actively on developing the technical means and mechanisms for combating terrorism. Those technical means may include the technologies and systems necessary for surveillance, monitoring and intelligence-gathering (such as sensors and the platforms in which these sensors would be embedded), for processing and disseminating this information (i.e. computing and communications), for coordinating counter-terrorist responses (such as command and control networks, encryption and so forth), and even for engaging in military or paramilitary counter-terrorist operations (i.e. less-than-lethal antipersonnel weapons, chemical biological radiological and nuclear (CBRN) containment and disposal systems, platforms and systems for patrolling borders and protecting critical infrastructures, and joint special operations).

Obviously, there are many potential incentives and opportunities for expanding cooperation in critical areas of science and technology (S&T) as it applies to developing, producing, and deploying systems for homeland defence and counter-terrorism. The potential economic, military, and political benefits accruing from such cooperation are readily apparent. They help spread the costs and reduce the risks of designing and developing next-generation technologies among several players, while simultaneously reducing duplicative research and development (R&D) efforts. They enhance standardisation and interoperability when it comes to joint counter-terrorist operations, and bolster international cooperation in general on anti-terrorism. They help increase economies of scale in production and theoretically make manufacturing processes more efficient. Finally, they facilitate innovative research by creating critical, transnational assemblages of know-how when it comes to science and technology.

At the moment, however, the United States, Western Europe, the former Soviet Union, and East Asia are pursuing their own security-related S&T programmes, often with little regard for the potential economic, technological, and national security benefits that could come from pooling resources and efforts. Moreover, many of these technologies will come from both inside but increasingly also outside of the traditional military technological-industrial complex. Accordingly, there is the potential for new international initiatives to reach out to non-traditional sources of expertise.

This Chapter will examine the opportunities for expanding international cooperation in the area of S&T when it comes to the challenges of homeland defence and combating transnational terror. It will also address current initiatives and challenges to collaborative S&T in these areas and lay out some suggestions for improving future cooperation.

2. Opportunities for International Cooperation on S&T for Homeland Defence and Counter-Terrorism: The RMA and Transatlantic Arms Collaboration

There exist several areas of potential cooperation on S&T for homeland security and counter-terrorism where the allies in the global war on terror have common requirements for improved capabilities. Many of these have critical overlap with requirements for enhancing "conventional" defence needs. As a recent European Union (EU) report on the future of European security-related research and technology programmes puts it:

> "[M]any capabilities serve internal and external as well as military and non-military security purposes. Surveillance, for example, is needed for both the protection of national borders and for crisis management operations abroad.

> The same is true for secured communications, intelligence and assessment capabilities" [2, p. 19].

Specific areas of potential S&T cooperation include:

- Sensors: CBRN scanning and detection devices; electro-optical, electronic, acoustic, infrared, thermal, millimetre-wave, radar, and laser surveillance and screening systems.

- Platforms (for sensors): satellites, unmanned aerial, underwater, and surface vehicles (UAVs/UUVs/USVs); manned systems, such as the Aerial Common Sensor or the Fuchs NBC (nuclear-biological-chemical) Reconnaissance System.[2]

- Information Technologies: interoperable command, control, and communications networks; encryption software; biometric passports; fingerprint or facial recognition software; specialised databases (such as the US-VISIT fingerprint registry and cross-reference project); secure data links; modelling and simulation.

- Counter-terrorist defences: CBRN containment and decontamination systems; missile protection systems for passenger jets; coast guard defences; missile defences.

With regard to increasing and enhancing transnational cooperation on such systems and technologies for counter-terrorism and homeland defence, it is important to recognise the potential spin-off benefits to be found in such unlikely areas as the so-called revolution in military affairs (RMA) and the already extensive efforts to expand international collaboration on conventional armaments. Both areas hold considerable promise for expanding international S&T cooperation when it comes to combating terrorism.

The emerging information technologies (IT)-based RMA is already having a transformative effect on armed forces around the world, and the IT-RMA is fundamentally altering modern visions of the character and conduct of warfare. For example, dramatic advances in information technologies over the past two or three decades – supplemented by recent advances in new materials and construction techniques – have made possible significant innovation and improvement in the fields of sensors, seekers, computing and communications, automation, operational range, and precision-strike. Central to the IT-led RMA is the emerging concept of network-centric warfare (NCW): warfare based on vastly improved battlefield knowledge and connectivity, due to technological breakthroughs creating more capable command, control, communications, computing, intelligence, surveillance, and reconnaissance (C4ISR) networks.

Additionally, the IT-based RMA is synergistic: it entails the integration and employment of C4ISR systems, platforms, and weapons (particularly "smart" munitions) in ways that increase their effectiveness and capabilities beyond their individual sets. This "bundling together" is reminiscent of Admiral William Owens's "system of systems" concept, in that it entails the linking together of several types of discrete and even disparate systems across a broad geographical, inter-service, and electronic spectrum in order to create new, so-called "core competencies" in war fighting.

Clearly, many of the technologies fundamental to the RMA can be readily applied to homeland security and counter-terrorism. In particular, NCW, with its emphasis on improved intelligence and sensing, communications, and command and control, is especially pertinent to the requirements of homeland defence and counter-terrorism [3, 2, p.12]. Indeed, the Bush administration's concepts of homeland security and counter-terrorism draw heavily on concepts and even the language of the RMA. For example, the White House's policy paper, *National Strategy for Homeland Security*, released in July 2002, acknowledges that "information and information technology," is something that "every government official performing every homeland security mission depends upon" [4, p.55]. The document specifically refers to the centrality of "scanners, sensors and monitors to collect data, which they would then store, disseminate and analyse using information technologies such as databases, fiber optic networks and software programs" [5, p.71]. The paper even expropriates from Admiral Owens's "system of systems" conceptualisation:

> "We will build a national environment that enables the sharing of essential homeland security information. We must build a 'system of systems' that can provide the right information at all the right times. Information will be shared 'horizontally' across each level of government and 'vertically' among federal, state and local governments, private industry, and citizens. With the proper use of people, processes and technology, homeland security officials throughout the United States can have complete and common awareness of threats and vulnerabilities... We will leverage America's leading-edge technology... to effectively secure the homeland" [4, p.56].

The growing trend in the internationalisation of conventional arms production is also a promising situation for expanding cooperation in developing new technologies for fighting terrorism. Many of these systems and technologies are dual-use in the sense that they can be employed for homeland defence as well as more conventional defences (for example, coast guards). The transnational development and manufacture of conventional armaments has grown significantly over the past several decades, and this globalisation process is dramatically transforming the world's defence industry [6]. Since the early 1960s, there have been scores of collaborative arms projects involving two or more countries, entailing the transfer or joint development of military technologies and combat systems, as well as the sharing of financial risks and rewards. Current transatlantic cooperative arms programmes include the NATO Airborne Ground Surveillance (AGS) system, the Evolved Sea Sparrow missile, the Multifunction Information Display System (MIDS), the Medium Extended Air Defence System (MEADS), and the F-35 Joint Strike Fighter. In Western Europe, nearly all armaments production is now collaborative (and major programmes include the Eurofighter jet, the Tiger attack helicopter, the Meteor air-to-air missile and the A400M transport aircraft). This process is increasingly rooted in joint-venture companies such as Eurocopter (a Franco-German helicopter producer), MBDA (an Anglo-French-Italian missile company), and Astrium (a multinational satellite firm). Finally, there is increasing international cooperation at the level of major subsystems, such as radars, and engines, and an already considerable transatlantic trade – $12 billion a year, by one estimate – in defence-related components and subsystems, such as computer chips, displays, flight control systems, and ejector seats [7, p.273].

There is considerable potential, therefore, for homeland security and counter-terrorism to benefit from both traditional defence and RMA-related research and technology programmes. Technologies for sensing and scanning, unmanned vehicles, standoff defence, command and control and so forth are equally applicable to homeland defence. It is readily conceivable that counter-terrorist S&T programmes could either piggyback on defence-related R&D or else utilise existing models of cooperation to fashion international cooperative S&T activities specifically for counter-terrorism.

3. Initiatives and Vehicles for International Cooperation on S&T for Counter-Terrorism

Europe has a long history of both combating terrorism and collaborating on innovative research and technology; it should be no surprise, therefore, that these two experiences should eventually coincide. Europe has long been both a target and a base for terrorist groups, both globally oriented – particularly Al-Qaida – but also "home grown," such as the Irish Republican Army and the Basque ETA. This terror threat is only heightened by the possibility of the further proliferation of weapons of mass destruction and the prospect that such weapons could end up in the hands of terrorists [8, pp.4-5]. Since terrorism is increasingly transnational, Europe – and particularly the EU – has increasingly recognised the need for a comprehensive and cooperative response to such threats. This includes pooling resources and technological strengths in order to wring maximum efficiency and effectiveness from its technological and industrial base. As a European Commission supported document, *Research for a Secure Europe*, put it:

> "Europe has high quality research institutes and a substantial and diverse industrial base from which to address technology requirements in the security domain. However, structural deficiencies at the institutional and political level hinder Europe in the exploitation of its scientific, technological and industrial strength. The dividing line between defence and civil research; the absence of specific frameworks for security research at the EU level; the limited cooperation between Member States and the lack of coordination among national and European efforts – all serve to exacerbate the lack of public research funding and present major obstacles to delivering cost-effective solutions.
>
> "To overcome these deficiencies, Europe needs to…improve the coherence of its efforts. This implies…effective coordination between national and European research activities…systematic analysis of security-related capability needs, from civil security to defence…full exploitation of synergies between defence, security and civil research…[and] institutional arrangements that are both efficient and flexible enough to combine Member State and Community efforts and to involve other interested partners" [2, p.6].

In this regard, therefore, past and present European experiences with collaborative defence-related S&T programmes can be a useful guide for expanding cooperation in the area of developing technologies for combating terrorism. Certainly armaments collaboration faces many of the same challenges: dissonant requirements, competing programmes, wasteful duplication of effort and so forth, and efforts to overcome these hurdles could provide important "lessons learned" for those trying to increase S&T cooperation when it comes to homeland security and counter-terrorism.

Europe, in fact, has long promoted the idea of collaborative defence-related S&T, and it has advanced this process through a variety of initiatives and programmes. EUCLID (European Long-Term Initiative in Defence) is an initiative created in the late 1980s by the Western European Armaments Group (WEAG), with the objective of strengthening

Europe's position in defence research and technology. The EUCLID Programme involves both industry and research institutes, and has established at least 120 specific "research and technology" (R&T) projects that are part of thirteen Common European Priority Areas (CEPAs), such as "advanced information processing and communication", "defence modelling and simulation technologies," and "missile, UAV, and robotic technologies". In 2002, EUCLID received approximately 100 million euros in funding [9].

EUROPA (European Understanding for Research Organisation, Programmes, and Activities) is an umbrella memorandum of understanding (MoU) established by WEAG in May 2001 to promote cooperative R&T research on defence projects. EUROPA has no formal guidelines, but rather encourages signatory nations to establish international teaming arrangements (known as European Research Groups, or ERGs) to engage in collaborative R&T research programmes. These ERGs, in turn, determine their own rules with regard to contracting, intellectual property rights, funding, and security [9].

Other WEAG MoUs covering pan-European S&T research are THALES (Technical Arrangements for Laboratories for Defence European Studies), SOCRATE (System Of Cooperation for Research and Technology in Europe), and the Memorandum of Understanding on Test Facilities. In addition, the so-called Framework Agreement (formerly the Letter of Intent, or LOI) among the six leading European arms manufacturers (the United Kingdom, France, Germany, Sweden, Italy, and Spain) establishes ground rules for European cooperation on defence-related research and development, such as improved information-sharing regarding defence-related R&T programmes, the harmonisation of requirements and export controls, security of supply and classified information.

Most recently, as laid out in *Research for a Secure Europe*, a high-level group has called for the European Union to forge a new European Security Research Programme (ESRP) to promote expanded pan-European S&T research in developing improved capabilities for homeland security, especially counter-terrorism. The ESRP initiative particularly calls for the "effective coordination between national and European research activities," and the "full exploitation of synergies between defence, security, and civil research," in order to "bridge the gap between civil and traditional defence research" and "maximize the benefits of multipurpose aspects of technologies" [2, pp. 6, 20-21]. The document argues that "in order to allow for a maximum of cross-sector interaction," an ESRP should in particular:

- Look at the "crossroads" between civil and defence applications;
- Foster the transformation of technologies across the civil, security and defence fields; and
- Define multi-functionality as positive criteria for the selection of research proposals [2, p.21].

The document also adds that:

> "An ESRP should concentrate on interoperability and connectivity as key functions for security management in a distributed environment. Emphasis should also be placed on security areas that require a particularly high degree of cross-border and inter-service cooperation. This is the case for measures against bio-terrorism, for example, where investments would bolster existing health and emergency infrastructures and thus be beneficial for society as a whole. In addition, architectural design rules for European efforts as well as common standards and protocols for 'systems-of-systems' should be defined at an early stage to enhance IT security and interoperability between different systems and user communities" [2, p.21].

In the United States, the Department of Homeland Security (DHS) is the centre for national efforts regarding domestic security and counter-terrorism. According to the DHS's website:

> "The Department of Homeland Security is committed to using our nation's scientific and technological resources to provide federal, state, and local officials with the technology and capabilities to protect the homeland. The focus is on catastrophic terrorism – threats to the security of our homeland that could result in large-scale loss of life and major economic impact. The Department's Science & Technology directorate works to counter those threats, both by evolutionary improvements to current technological capabilities and development of revolutionary, new technological capabilities" [10].

In FY2004, the DHS directly managed a budget of $36 billion and coordinated an overall budget exceeding $100 billion distributed among various other departments and in individual states. In particular, the DHS budget included around $1 billion devoted to research, and it has established its own science and technology programme specifically for homeland defence and counter-terrorism. DHS science and technology activities are in addition to those funded by other agencies also engaged in homeland defence and particularly defence R&D programmes funded by the U.S. Department of Defense (DoD) that could be applicable to homeland security.

Compared to Europe, international defence-related S&T cooperation involving the United States has generally been much more *ad hoc*. The United States has collaborated with Germany on the X-31 experimental aircraft and with the United Kingdom on advanced vertical takeoff technologies. Japan and the United States together have instituted several major defence-related research projects – in particular, joint research on ducted rocket engines and next-generation missile defence technologies. Given the low profile of most basic research programmes, it is difficult to arrive at a reliable figure for how many of these projects are currently underway and how many of these may have a potential application for homeland defence. Nevertheless, given the U.S. government's strong support for foreign participation in such critical defence programmes as the JSF, the Deepwater U.S. Coast Guard recapitalisation programme[3] and missile defence, and for widening the global war on terror, it would seem reasonable to conclude that the United States is open to the idea of expanding transatlantic and international collaboration on S&T development for counter-terrorism. This view is affirmed by the Chapter by Dockery and Albright in this volume that sets out the DHS stance on international S&T cooperation.

4. Barriers to International Cooperation

There are many incentives and opportunities for countries to pool their efforts when it comes to counter-terrorist-related S&T activities. Nevertheless, even at the level of research, there remain strong national protectionist and parochial barriers to increasing such collaboration.

Such impulses and measures have long hobbled international collaboration in the area of conventional armaments. In the United States, "Buy America" laws or bureaucratic biases have frequently shut foreign firms out of potentially lucrative U.S. defence contracts. In late 2003, for example, the U.S. House of Representatives sought to insert into the FY2004 defense bill a provision that would have raised the threshold for U.S. content and furthermore would have applied this rule to civilian technologies used for arms production, such as machine tools [11, 12]. Specific to homeland defence and counter-terrorism, the U.S. Homeland Security Act of 2002 restricts the DHS's ability to award contracts to foreign-operated corporations, including U.S. subsidiaries of companies incorporated in other countries; for example, DHS was criticised for awarding a contract to Accenture LLP, a U.S.-based firm owned by a company chartered in Bermuda, to implement the Department's US-VISIT program. [13, pp.1, 4].

Within the Department of Defense (DOD), meanwhile, it is argued that a "default mode" favouring American-made defence technologies and systems still pervades the Pentagon bureaucratic culture when it comes to arms procurement, and that the DOD still views foreign participation in U.S. acquisition to be a nuisance at best [14, p.26]. Finally, highly restrictive U.S. export controls and technology transfer regulations are also frequently criticised for hindering international collaboration on defence and homeland security; export licenses, for example, are generally required even for technical discussions with foreign nationals.

Protectionism and parochialism are hardly the sole purview of the United States, however, and many European countries have admittedly had their own problems with committing to international – and particularly transatlantic – S&T cooperation. Europe as a whole has often been lukewarm to collaborative projects that place it in a decidedly junior role, even though it may lack the funding or the technology to participate at a higher level. Consequently, a "European preference" has been increasingly evident in recent procurement decisions made by European governments, such as the A400M transport aircraft, the Galileo navigation satellite project, and the Meteor missile [14, p.21]. Even within Europe, S&T cooperation on defence and homeland security is modest compared to domestic activities. Notwithstanding initiatives such as EUCLID and EUROPA, most European nations continue to emphasise national R&T programmes, and most funding for defence-related research remains inside each country.

One hopeful sign for expanding and enhancing transatlantic cooperation on homeland security and counter-terrorism can be found in the recent European Commission sponsored report, *Research for a Secure Europe*. It recognises that the United States, due to the large scale and scope of its investments in homeland defence, will naturally dominate R&D in this area:

- The U.S. is taking a lead and will develop technologies and equipment which, subject always to U.S. technology transfer permission, could meet a number of Europe's needs;
- U.S. technology will progressively impose normative and operational standards worldwide;
- In certain areas, where the U.S. authorities prioritise their investment and achieve fast product "speed to market," U.S. industry will enjoy a very strong competitive position [2, p.21].

Consequently, it concedes that

> "Europe's response to these developments will need to be realistic. On one hand, global interoperability requires universal solutions: for example, a system seeking to track and control the international movement of freight containers will have to comply with regulations for containers destined for the U.S., which alone account for 50% of international container traffic. Similarly, there is limited value in duplicating research already conducted elsewhere if the results can be shared in a mutually beneficial way.

Furthermore, the evaluation of the case for investment in Europe will need to take into account the dynamics of the market" [2, p. 21].

Therefore, while the report argues that, "for critical technologies, Europe should aim for an indigenous competitive capability, even if this involves duplication of effort," it agrees that, "for less critical technologies and/or in areas where requirements in Europe are distinct or in advance of those sought elsewhere," consideration should be given to cooperative development or even off-the-shelf buys [2, p. 21].

Another positive sign is a recent decision by the U.S. Department of Homeland Security to assign a full-time official to the EU, to help coordinate transatlantic cooperation on standardisation, harmonisation, and reciprocity in various homeland defence measures. One of the first of these initiatives will be developing common standards for data for biometric-enhanced passports [15].

5. Conclusions

There are many incentives and opportunities for expanding international cooperation on S&T for homeland defence and counter-terrorism. Certainly the allies in the global war on terror have many reasons for joining forces and working together in directly combating terrorism. In many ways, the aims and objectives in such collaboration – both broadly in an operational sense, and more narrowly in a sense of cooperating on developing technologies and capabilities to better fight the war on terror – are clearer and more focused than they were during the Cold War. Certainly the threat is more tangible, if more limited, and therefore the requirements of such collaboration are, realistically speaking, more easily attainable.

Despite this recognition, the allies in the global war on terror still have their work cut out if they are to overcome the longstanding barriers and impediments that have traditionally conspired to make international collaboration difficult. There are many measures and initiatives that the leading nations in this new war could consider to improve and enhance cooperation on S&T for homeland security and counter-terrorism, including:

- International harmonisation of requirements and acquisition processes for homeland defence;
- Increased overall funding for homeland defence-related S&T among the leading nations combating terrorism;
- Establishment by the DHS of a programme specifically designed to underwrite collaborative S&T;
- Concurrently, the expansion of European S&T initiatives, such as EUCLID and EUROPA, to include countries and firms outside of Europe;
- International agreement on security of supply and information; and
- Reform of export controls.

Many of these proposals have also been advanced, in slightly different forms, for the purposes of expanding international arms collaboration, so much of the conceptual groundwork has already been done, and homeland security-related S&T collaboration can piggyback on these endeavours. Above all, however, collaborating nations must not permit protectionist and parochial sentiments to obstruct progress on crucial international efforts to acquire the capabilities to combat terrorism.

There are some hopeful signs. At the same time, in formulating new initiatives and programmes for cooperative transnational S&T research for counter-terrorism, proponents have an important advantage over more complicated, weapons-related research and development projects. Science and technology development is usually seen as research, and research is by its nature more experimental rather than developmental, more modest in scope and ambition, and relatively low-key and low-level in execution. As such, when it comes to directing international defence- or counter-terrorist-related S&T research, there are distinct advantages over full-scale R&D activities (such as large-scale efforts like the Joint Strike Fighter or the Meteor missile). Such research is generally much less expensive to conduct; the organisational arrangements are usually looser and more informal (e.g., strategic alliances, as opposed to consortia), and therefore project management is relatively less complicated; there are fewer problems when it comes intellectual property rights (technological breakthroughs tend to be shared equally at this stage); and politically it is easier to enter into and engage in, as there are fewer economic benefits to fight over at that stage (such as jobs and production rights) and fewer adverse political repercussions if a partner decides to withdraw from a project.

Notes

1 The analyses and opinions expressed in this paper are strictly those of the author and should not be construed as representing those of the U.S. Department of Defense or of any other U.S. government organisation.

2 The Aerial Common Sensor is a U.S. Army airborne surveillance and reconnaissance system equipped for imagery and signals intelligence-collection. The Fuchs is a German-developed armoured reconnaissance vehicle equipped with systems for detecting, measuring, and evaluating potential NBC agents.

3 The Deepwater program is a multifaceted plan to outfit the U.S. Coast Guard with new surface vessels, helicopters, patrol aircraft, and UAVs, and to link their operations with a state-of-the-art C4ISR network. Several European companies, including the UK's BAE Systems (communications systems), Italy's Agusta (helicopters), and Sweden's Karlskrona (ship materials technologies), are partnering in this programme.

References

[1] Borgu A. Combating Terrorism – The Challenge for Policymakers in Australia and the U.S. [conference paper]. Brisbane (Australia); 2004.
[2] Group of Personalities in the Field of Security Research. Research for a Secure Europe. Luxembourg: European Commission; 2004.
[3] Mulholland D. Homeland defense market grows. Jane's Defence Weekly. August 25, 2004 (Internet version). Available at: http://jdw.janes.com/public/jdw/index.shtml
[4] Office of Homeland Security (US). National Strategy for Homeland Security. Washington (DC): Office of Homeland Security; July 2002. Available at: http://www.whitehouse.gov/homeland/book/nat_strat_hls.pdf.
[5] Harris S. Disconnect. Government Executive: 2002; 34: 71-80.
[6] Bitzinger RA. Towards a Brave New Arms Industry? London: International Institute for Strategic Studies/Oxford University Press; 2003.
[7] International Institute for Strategic Studies (IISS). The Military Balance 1998-1999. Oxford: Oxford University Press; 1999.
[8] Solana J. A Secure Europe in a Better World. Thessaloniki: European Council; 2003.
[9] Western European Armaments Organization website (http://www.weao.weu.int).
[10] U.S. Department of Homeland Security website (http://www.dhs.gov).
[11] Mulholland D. US Industry: bucking the trend? Jane's Defence Weekly; February 25 2004 (Internet version). Available at: http://jdw.janes.com/public/jdw/index.shtml
[12] Muradian V, Matthews W. White House shifting on Buy-American? Defense News; September 30 2003 (Internet version). Available at: http://www.defensenews.com/
[13] Carafano J. Protectionism Compromises America's Homeland Security. Washington, DC: Heritage Foundation; June 9 2004.

[14] Center for Strategic and International Studies (CSIS). The Future of the Transatlantic Defense Community: Final Report of the CSIS Commission on Transatlantic Security and Industrial Cooperation in the Twenty-first Century. Washington, DC: CSIS; January 2003.

[15] Tigner B. U.S., EU to improve counterterror cooperation: Ridge. Defense News. January 13, 2005 (internet version). Available at: http://www.defensenews.com/

Part 5

The Governance of Science and Technology in the New Security Environment

Science and Technology Policies for the Anti-Terrorism Era
Andrew D. James (Ed.)
IOS Press, 2006

Impacts of Post-September 11 Security Policies on U.S. Science

Albert H. TEICH[1]
Director, Science & Policy Programs
American Association for the Advancement of Science
1200 New York Ave., NW, Washington, DC 20005 USA

Abstract. The U.S. has passed a number of new laws and tightened enforcement of others since the attacks of 9-11. These laws are affecting scientific research and higher education in many ways. This Chapter explores the impacts on science in several areas: foreign students and visitors; the use of certain chemical and biological agents in the laboratory; restrictions on federal grants and contracts; publication of research results, and dissemination of "sensitive but unclassified" scientific information. Some of the new laws and policies may turn out to be counterproductive, reducing U.S. security by retarding scientific and technological development.

1. Introduction

In the wake of the attacks on the World Trade Center and the Pentagon on September 11, 2001 and the anthrax-laced letters that were sent to at least two Senators and a number of media personalities soon after, the U.S. Congress quickly passed several pieces of legislation intended to counter the threat of terrorism, reinforce border security, and strengthen the hand of U.S. law enforcement agencies. Federal agencies also toughened the enforcement of existing laws and regulatory policies. The new laws and enforcement efforts have affected life in the United States in many ways, from increased security at airports and public buildings to periodic yellow, orange, and red alerts, to restrictions on civil liberties. Scientific research and higher education have been affected more strongly than many other areas and often in ways not anticipated or intended by the framers of the new laws and policies.

The impacts on research and education are being felt in a number of ways: (a) declines in the number of foreign students and scholars coming to the U.S., due largely to delays in processing and denials of visa applications; (b) new regulations requiring stringent controls over laboratory use of a wide range of chemical and biological substances (known as "select agents"); (c) the appearance of restrictive clauses in federal research grants and contracts to universities requiring pre-publication review of research results or requiring export licenses for the involvement of non-U.S. citizens in the research; (d) growing concerns about publication of basic research results (especially in the biosciences) that could be employed in weapons by terrorists or rogue states; and (e) efforts by some government agencies to control the dissemination of "sensitive but unclassified" information.

2. Foreign Students and Visitors: Visa Issues

2.1 New Legislation and Procedures

When the USA PATRIOT Act of 2001 and the Enhanced Border Security and Visa Entry Reform Act of 2002 tightened both the requirements and the enforcement of entry procedures for foreign visitors, academia was one of the areas affected most immediately.

The USA PATRIOT Act, the name of which is a somewhat tortured acronym for "Uniting and Strengthening America by Providing Appropriate Tools Required to Intercept and Obstruct Terrorism," is perhaps the most controversial of the several pieces of legislation passed in the wake of September 11. It was enacted only 45 days after the terrorist attacks, with no Congressional hearings and little discussion.

The act has four main thrusts: firstly, it strengthens the hand of law enforcement agencies in using wiretaps and other means of intercepting electronic communications; secondly, it eases restrictions on foreign intelligence gathering within the United States; thirdly, it expands the government's authority to regulate the activities of U.S. financial institutions in order to prevent money laundering; and fourthly, it includes a number of provisions aimed at preventing terrorists from entering the U.S. as well as facilitating the detention and deportation of alien terrorists and others who support them. Under the last of these headings are several elements that affect immigration, including broadening the grounds for excluding individuals from the U.S. and expediting the implementation of a previously authorised entry and exit data system for all visitors as well as a foreign student tracking system authorised in 1996 but up to that time not implemented.

The Enhanced Border Security and Visa Entry Reform Act of 2002 was signed into law by President Bush a few months after the USA PATRIOT Act, on May 14, 2002. Provisions of this act include enhanced review of visa applicants, improved coordination among federal agencies engaged in immigration-related activities and strengthening elements of the previously-enacted foreign student monitoring system. The act requires consular officers issuing visas to obtain electronic verification of a student's admission from the institution before issuing a visa and also requires the officer to obtain names and contact information for family members who can verify information regarding applicants.

New procedures implemented subsequently require personal interviews for all non-immigrant visa applicants. The number of circumstances under which this requirement can be waived by a consular officer has been reduced. The interview requirement poses a substantial hurdle for foreign students applying for U.S. visas, because it requires the applicant to travel to the nearest consulate (which can be prohibitively expensive for some students) and because the backlog of applicant interviews has led to lengthy delays in scheduling interviews and, therefore, in visa processing.

2.2 Impacts on Foreign Students

Anecdotal reports of visa problems experienced by students and current and prospective visiting scholars began to surface in the months after September 11. More recently, surveys have shown significant declines in applications, acceptances, and enrollment of foreign students in U.S. colleges and universities, especially at the graduate level [1]. For over ten years prior to September 11, international graduate student enrollment had shown steady growth. According to surveys conducted annually by the Council of Graduate Schools, that trend has reversed. Between 2003 and 2004, first-time international graduate student enrollment at U.S. institutions declined by 6 percent. The decline in enrollment follows similar reported declines in applications and admissions [2].

Somewhat surprisingly, the countries hit hardest by the tighter entry requirements have been China and India. Graduate applications from students in China declined by 45 percent between 2003 and 2004 and those from India fell 28 percent in the same period [2]. Graduate enrollment of Chinese and Indian students has also been down, though not quite as much. These trends are noteworthy because (a) neither of these countries is regarded as posing a terrorist threat and (b) students from these countries constitute a significant share of graduate students (as well as teaching and research assistants) in science and engineering fields. In general, it is reported that most visa delays and denials among these students are due not to security concerns but rather to the students' inability to demonstrate their intent to return home after completing their studies.

Also affecting enrollments in scientific and engineering fields is the "Visas Mantis" programme - a programme designed to protect against the transfer of sensitive technologies, generally defined as those subjects on the government's Technology Alert List. The programme requires greater scrutiny of applications from students who wish to study these subjects. While the list includes such obviously sensitive subjects as enrichment of fissile material and ballistic missile systems, it also covers subjects that, while they might be considered dual-use, are not usually viewed as posing threats to American interests, such as immunology, artificial intelligence, and landscape architecture.

After September 11, the number of visa applicants reviewed under the Visas Mantis programme grew substantially. During 2000, some 1,000 non-immigrant visa applications were flagged for review; by 2002, that number had risen to 14,000. Government agencies involved in the review process were not prepared for this growth. In a study of 71 cases published in February 2004, the Government Accountability Office (GAO, formerly the General Accounting Office) found that it took an average of 67 days to review applications requiring the Visas Mantis process [3]. With such long delays, students are becoming increasingly discouraged about coming to the U.S., which may help to explain increases in foreign applicants to universities in other Anglophone countries. Australia, for example, reportedly experienced a 16.5 percent increase in the number of foreign students in the 2003 academic year [4].

Increased security has also kept many foreign students and scholars from attending scientific meetings in the U.S. In November 2003, two Chinese-born University of Toronto students were prevented from attending a scientific conference in Austin, Texas, despite the fact that they had applied for visas, booked all travel plans, were invited to present research posters and had attended the same meeting in 2002 [5]. It took 3 months to process the security and background checks on the students, which was well after the end of the conference. A year earlier, in October 2002, a University of Toronto professor, a Canadian citizen born in Iran, was so angered by having to be fingerprinted, photographed and questioned prior to coming to chair a National Science Foundation meeting in the U.S., that he decided to cancel his trip [6]. Such rigorous background checks of Canadian citizens born in Iran, Iraq, Libya, Sudan or Syria were part of the National Security Entry Exit Registration System (NSEERS) law that went into effect nationally on October 1, 2002 [5]. NSEERS has since been replaced by US-VISIT and SEVIS.

US-VISIT is a new set of entry procedures that applies to all nonimmigrant visitors (including students, tourists and people on business) regardless of their country of citizenship. It went into effect at 115 airports and 14 seaports in January 2004 and is scheduled to be expanded. The procedures involve having a visitor's two index fingers scanned and a digital photograph taken at the port of entry. SEVIS (Student and Exchange Visitor Information System) is a computer database system operated by the federal government that manages data about foreign students and exchange visitors while they are in the U.S. It replaces an outdated paper-based system that had been in existence for many years. SEVIS went into operation in August 2003 and participation is now mandatory for all foreign student and exchange visitor visa applicants and the institutions that host them.

The existence of the SEVIS database and the $100 fee that visa applicants must pay to register in it may be another factor contributing to the perception among foreign students and visitors that the U.S. government is erecting barriers to keep them out - a perception that universities and government officials are trying hard to dispel. Nevertheless, the process is now routinised and by itself does not seem to be a major driver of the decline in foreign student enrollments in U.S. universities. Of greater consequence have been the increases in visa delays and denials that prospective students have faced, as discussed above.

2.3 Response of the Research Community and Subsequent Developments

Believing that improvements to the visa process need to occur soon and can be made without jeopardising the security of the nation, many scientific, academic and professional organisations have been urging reforms. In May 2004, 25 leading scientific, engineering and educational organisations representing nearly 95 percent of the academic research community in the U.S., sent a statement to President Bush and Congress with six suggestions on how to improve the current visa process [7]. The suggestions included:

- Extending the validity of Visas Mantis security clearances for international students and scholars from the current one-year time period to the duration of their course of study;
- Establishing more timely visa renewal processes including renewals that can be started prior to leaving the U.S.;
- Creating a mechanism whereby applicants and sponsors can track their applications and have applications pending for 30+ days moved to the top of the waiting list;
- Ensuring clear protocols and trained consular staff for screening applications so as to avoid inconsistent treatment;
- Revising visa reciprocity agreements between the U.S. and countries like China and Russia that send large numbers of scholars and students in order to reduce the number of times visiting scholars have to renew their visas;
- Implementing a fee collection system for SEVIS [7].

Federal government officials have responded favourably to a number of these recommendations. Discussions are underway among several agencies regarding the possibility of extending the duration of Visas Mantis clearances, although the issue has not yet been resolved. The State Department has, however, been working to speed up the processing of Visas Mantis applications. According to department officials, average processing time was two months in 2003. By late 2004, 98 percent of all Visas Mantis cases were being processed in less than 30 days and 85 percent were cleared in less than three weeks. This, together with an increase in the number of consular positions, and improved training for Foreign Service Officers, has helped to reduce substantially the backlog of visa applications. In addition, the State Department has begun posting average visa appointment wait times at various U.S. embassies and consulates on the Internet and is giving priority to students and scholars in scheduling interviews.

The main problem now, officials stress, is to make these improvements better known so that prospective students will not be discouraged from applying by impressions of what the situation was like previously. To that end, State Department officers, from the Secretary of State on down, have been giving interviews and speeches and writing editorials to let it be known that the U.S. welcomes foreign students and visitors [8].

3. Research in Biological and Chemical Laboratories

3.1 New Laws Affecting Research

Two laws passed in the wake of the terrorist attacks of September 11, 2001, and the anthrax letters that followed in October have dramatically altered the climate for research with pathogens in the United States. Scientists working with "select agents," and their institutions, must comply with these laws or face civil and/or criminal penalties.[2]

The USA PATRIOT Act sets criminal penalties for possessing a biological agent or toxin for purposes other than "bona fide research," and requires that "no restricted person . . . shall . . . transport . . . or possess . . . any biological agent or toxin, or receive any biological agent or toxin" listed as a select agent.[3] The Public Health Security and Bioterrorism Preparedness Act of 2002 expands the definition of "restricted persons," and requires research institutions to submit the names of persons with access to select agents to the Department of Justice to determine if they fit that definition. Subsequent regulations assigned the authority for establishing and updating the select agents list to the Department of Agriculture's Animal and Plant Health Inspection Service for agents posing a threat to animal or plant health or products, and to the Centers for Disease Control and Prevention (an agency of the Department of Health and Human Services) for those agents that could potentially threaten public health and safety. The law requires facilities and individuals to register with these agencies "the possession, use, and transfer of listed agents and toxins", which now number more than 80.

3.2 Impacts on Researchers

These new hazardous agent rules have affected many scientists working on bacteria or viruses on the select agents list. Researchers using select agents in their work must now undergo time-consuming FBI clearances, including fingerprinting, criminal background checks, and interviewing of relatives and friends [10]. Some institutions have decided to forego all research involving select agents. The Centers for Disease Control and Prevention had expected 817 entities to register under the rules. By the initial deadline, however, only 323 had actually done so, suggesting that many have chosen to discontinue select agent research [11]. Researchers at Stanford chose to give up their research on select agents, believing that the "administrative and security burdens of the select agent rule outweighed the scientific need to maintain stocks on campus" [11]. For those institutions that choose to continue their research, the costs of upgrading facilities to conform to the new rules can be substantial. Louisiana State University at Baton Rouge reported spending approximately $130,000 in security systems for its labs, and the University of Louisville altered floor plans in its new science buildings in order to have offices and labs in separate places [12, 13]. Universities across the nation are feeling the economic strain accompanying these laboratory upgrades, and there are concerns that "smaller universities, without substantial financial resources, will be 'locked out' by regulations" [12].

3.3 Enforcement: Some Troubling Examples

Three well-publicised cases have made it clear that the government intends to enforce the new laws vigorously. Worth noting, however, is that none of the cases to date has had any connection to potential terrorists or terrorism.[4]

The most widely reported case is that of Professor Thomas Butler of Texas Tech University in Lubbock [14]. On March 10, 2004, Butler, a well-respected authority on infectious diseases, was sentenced to two years in prison, fined $15,000, and ordered to pay $38,000 restitution to his university, following an FBI investigation and trial that began when he reported 30 vials of plague bacteria missing from his lab in January 2003. Exactly what happened to the missing vials is still not known. The day after he reported them missing, Butler signed a statement saying that he had accidentally destroyed the vials. Later, he recanted the statement, claiming he had no idea what happened to them. Although Butler was cleared of the most serious charges against him which included lying to the FBI about the 30 missing plague vials, he was found guilty of illegally shipping dangerous pathogens and of other charges related to fees he had received from pharmaceutical companies for participation in clinical trials. Butler has appealed his conviction and the U.S. Department of Justice, in response, has appealed his sentence, claiming it was too lenient. Both appeals are pending as of early 2005 [15].

A few months before Butler's arrest, Tomas Foral had become the first person charged with mishandling a select agent under the USA PATRIOT Act [16]. Foral, a master's degree student at the University of Connecticut, was helping a professor clean out a laboratory freezer when he came across a container with some vials of anthrax-tainted cow tissue. Not knowing what to do with them, he put two in another freezer for possible future research. Although Foral insists the affair was strictly a misunderstanding, federal prosecutors charged him with unlawful possession of a select agent in violation of the new law - an offence punishable by up to ten years in prison. Eventually, Foral and the government reached an agreement that enabled him to avoid prosecution by doing two years of community service.

More recently, in May 2004, the indictment of Steven Kurtz, an art professor at the State University at Buffalo and Robert Ferrell, head of the Department of Human Genetics at the University of Pittsburgh School of Public Health, on two counts each of mail fraud and wire fraud, has sent chills down the spines of many in the scientific community. Ferrell allegedly gave Kurtz two strains of nonpathogenic bacteria which Kurtz intended to use in an art exhibit. Since the bacteria are considered harmless and are not listed as select agents, the Department of Justice is apparently building its case around the allegation that Ferrell violated his contract (material transfer agreement) with the American Type Culture Collection, from which he obtained the bacteria, by giving them to Kurtz. As of early January 2005, Kurtz is free on bond awaiting trial. Ferrell has been seriously ill and has not yet been arraigned [17].

3.4 Response

Reflecting growing concerns about the potential for misuse of biological research and the problems of preventing misuse without impeding scientific progress, a committee of the National Academies/National Research Council issued a report in October 2003 entitled *Biotechnology Research in an Age of Terrorism: Confronting the Dual Use Dilemma* [18]. One of the report's key recommendations was that the federal government establish an ongoing committee on biosecurity policy. Several months later, in March 2004, Secretary Tommy Thompson of the Department of Health and Human Services announced the creation of a National Science Advisory Board for Biosecurity (NSABB). The Board will consist of up to 25 voting members representing at least 15 governmental agencies. It will meet on a quarterly basis, more often if needed. The mission of the Board, which has been chartered initially for two years, is to:

- Advise on strategies for local and federal biosecurity oversight for all federally funded or supported life sciences research;
- Advise on the development of guidelines for biosecurity oversight of life sciences research and provide ongoing evaluation and modification of these guidelines as needed;
- Advise on strategies to work with journal editors and other stakeholders to ensure the development of guidelines for the publication, public presentation, and public communication of potentially sensitive life sciences research;
- Advise on the development of guidelines for mandatory programmes for education and training in biosecurity issues for all life scientists and laboratory workers at federally-funded institutions;
- Provide guidance on the development of a code of conduct for life scientists and laboratory workers that can be adopted by federal agencies as well as professional organisations and institutions engaged in the performance of life sciences research domestically and internationally [19].

4. Federal Research Funding to Universities and Government Security Policies

Another issue that has arisen in the current atmosphere of heightened security is the reported appearance of restrictive clauses in some federal research grants and contracts to academic institutions. The proposed clauses have generally involved requirements for pre-publication review of research products by the federal agency sponsors and background checks on foreign researchers participating in the research. In a system that has long been based on principles of scientific openness and the marriage of research and education, the appearance of such clauses in grant and contract negotiations for basic research projects has caused some consternation in the research community. The clauses have often involved the application of the concept of "sensitive but unclassified" to fundamental research and the concept of "deemed exports" to foreign nationals involved in such research. "Deemed exports" are discussed later in this section; "sensitive but unclassified" information is discussed in section 6.

4.1 Survey of "Troublesome Clauses"

Concerned about anecdotal reports from universities about the increasing number of such clauses appearing in grants and contracts, the Association of American Universities (AAU) and the Council on Governmental Relations (COGR), Washington-based organisations that represent the interests of research universities, decided to take a systematic look at the issue. Beginning in the spring of 2003, the organisations began to collect information from their member institutions on the occurrence of what they termed "troublesome clauses" in proposed federal contracts and grants received by these member institutions over a six month period. Twenty universities, representing both public and private institutions of varying sizes and regions of the country, were included in the study. The task force reviewed 138 reports of proposed awards that included clauses restricting publication or affecting participation of foreign nationals in research. The results of the survey are noteworthy and have been discussed extensively among university and government officials in Washington [20].

One hundred and five instances were reported in which universities received proposed funding instruments that contained restrictions on publication. There were 29 reported instances of restrictions on participation of foreign nationals; and four cases in which other kinds of access or dissemination restrictions were proposed by the funding agency. Many of these cases involved, not surprisingly, Defense Department (DOD) funding, either directly or indirectly, through subcontracts from DOD-funded industrial firms to universities. Other federal agencies were also involved, however, including the National Science Foundation, the Department of Health and Human Services, and even the Federal Highway Administration.

Universities responded differently to these situations. In 29 out of the 105 cases reported during the six month study period, the schools involved accepted the grants or contracts as proposed. In 46 cases they were able to negotiate alternative language which both they and the government found acceptable, while in nine cases the universities and the government were unable to come to terms and the universities rejected the awards. The remaining cases were still pending at the conclusion of the study period.

Findings were similar with regard to clauses involving the participation of foreign nationals. Thirty cases were reported. In ten, the university accepted the proposed clauses; in eleven they were able to negotiate revised language; and in four instances the contracts or grants were rejected. The remainder had not yet been settled.

In its recommendations, the AAU/COGR report stressed the need for federal research sponsors to recognise the distinction between academic, basic research, where openness and free exchange of information is essential, and research conducted under government sponsorship in industry, where traditions and practices are different [20]. Nevertheless, some policymakers seem to be moving in the opposite direction.

4.2 Reports from the Inspectors General on "Deemed Exports"

On projects at universities that it considers "sensitive but unclassified" - whether or not the institution regards them as basic research - the federal government has sought to expand its authority to intervene and increase restrictions. One example of such an intervention is requiring a lab employing foreign researchers to obtain a special license from the Department of Commerce under the department's Export Administration Regulations (EAR) in order to allow the researchers to work on certain projects or to use certain equipment [21]. The Export Administration Regulations, which have been in effect for many years, prohibit export of dual-use technologies on the Commerce Control List without a license. The regulations apply to information as well as physical goods, employing the concept of a "deemed export" - the transfer of information to a foreign national. Until recently, however, the concept has not often been applied to basic research in universities. Some government officials would like to change this.

In March 2004 the Inspectors General of the Defense Department (DOD) and the Department of Commerce (DOC) both issued reports reevaluating the application of export controls and security for foreign nationals in universities working with sensitive technology. The DOD report called for increasing incorporation of export control clauses in DOD contracts as well as expanding access control measures such as requiring badges for all foreign nationals and segregating controlled technology labs [22].

The DOC report focused on the use of sensitive scientific equipment controlled by the Export Administration Regulations (EAR). The report advocated increasing the application of "deemed export" provisions on all EAR-controlled equipment, even including equipment at universities involved in fundamental research [23]. AAU, COGR and other organisations representing university researchers, have voiced their concerns with these recommendations, citing examples of how tightening deemed export rules and restricting use of laboratory equipment by foreign nationals will "have a chilling effect on university research and education as well as compel discriminatory treatment of foreign nationals on campus" [24]. At the time of writing, the situation remains unresolved.

5. Scientific Publication

5.1 Problematic Papers?

In February 2001, Australian researchers published a paper in the *Journal of Virology* describing an experiment in which they inserted a gene called interleukin-4 into the mousepox virus, which is a relative of the smallpox virus. The researchers were looking for a way to stimulate the mouse's immune system to prevent reproduction. Instead, they found that the new gene made the virus more virulent. In fact, it killed mice that otherwise were immune to mousepox, including those that had been vaccinated. The paper caused something of a stir among scientists. Some were critical of the journal for publishing it, suggesting that it could tell potential terrorists how to create a smallpox virus that could circumvent existing vaccines. Others suggested that publication of the paper served the public interest by pointing out that authorities cannot rely only upon current vaccines to defend against smallpox; highlighting the need for more research on antiviral drugs; and, underlining the importance of quarantine in countering a smallpox epidemic [25].

The controversy over the mousepox paper and others with potential security implications spilled over beyond the scientific community into the political realm and took on new urgency after September 11 and, especially, after the anthrax attacks that followed, beginning a week later. Indeed at almost the same time as the anthrax attacks occurred, the full genome of *Yersinia pestis*, the bacteria that causes bubonic and pneumonic plague was published in the journal *Nature* [26]. A few months later, in May 2002, the *Proceedings of the National Academy of Sciences* published a study that showed how the smallpox virus uses a protein to evade the human immune system. An editorial that accompanied the paper noted that the journal had been advised by some people not to publish it on the grounds that it could provide assistance to terrorists [25].

Then, in the summer of 2002, *Science* published a paper on the synthesis of the polio virus from off-the-shelf materials. Even though the principle had been known for years (though never demonstrated experimentally) and it was known that the polio virus would not make a very useful bioweapon, Representative Dave Weldon, a Republican Congressman from Florida, expressed outrage. Weldon described the paper as "a blueprint that could conceivably enable terrorists to inexpensively create human pathogens for release on the people of the United States" [25]. He introduced a resolution in Congress criticising the journal and the American Association for the Advancement of Science, which publishes it, and calling on the federal government to review its policies relating to the publication of federally-funded research. Although the resolution did not come to a vote, the lesson - that policymakers would no longer ignore scientific papers if they seemed to have potential security implications - was not lost on the scientific community.

Nevertheless, it is not always easy to tell what research might be perceived as a threat. For example, a George Mason University graduate student's dissertation came under scrutiny in July 2003 despite the fact that it only used information that was publicly available. According to a report in *The Washington Post*, the student "mapped every business and industrial sector in the American economy, layering on top the fiber-optic network that connects them," thus laying bare the "country's nervous system" [27].

5.2 Beneficial Knowledge or Aid to Terrorists?

A major challenge in the debate over security versus openness in scientific publishing is the dual-use nature of much contemporary research, where research conducted for the advancement of knowledge or to serve completely legitimate ends can also be used as weapons [28]. While most scientists regard increasing knowledge as intrinsically a good thing, others disagree. Bioethicist Arthur Caplan, for example, worries that, "We have to get away from the ethos that knowledge is good. . . . Information will kill us in the techno-terrorist age." Furthermore, scientists cannot possibly develop defences against all conceivable biological warfare threats [28], and publication of what scientists do know could expose the nation's strengths and weaknesses - something which itself could be either good or bad. This is at the heart of the controversy surrounding the George Mason University graduate student's dissertation, which has been called "a cookbook of how to exploit the vulnerabilities of our nation's infrastructure" [27].

These discussions are reminiscent of the debates in the nuclear physics community that took place around the outset of World War II or those of the 1980s among computer scientists and mathematicians regarding potential misuse of cryptographic research. Advocates of openness contend that restrictions on publication would prevent researchers from pursuing important and useful studies that could benefit society and that, in any case, they would not halt, but only delay, the spread of information, since, if one researcher or group can make a discovery, so can others [28]. Others speak of copies of scientific articles and downloads from the Internet found in al-Qaeda caves in Afghanistan as evidence that terrorists actively seek scientific information for nefarious purposes. They counter that, at the very least, one should not simply hand such potentially dangerous information to them. The Bush Administration seems to agree. In early 2002, a White House spokesman was quoted in *The New York Times* suggesting that scientific journal editors not publish "sections of articles that give experimental details researchers from other labs would need to replicate the claimed results" to create biological weapons [29].

5.3 Response of the Scientific Community: Self-Regulation

In response to policymakers' suggestions that government might impose restrictions on scientific publication, publishers of scientific journals, especially those in the biotechnology field, have proposed various forms of self-regulation - developing their own statements of principles and review procedures to alert editors to sensitive information. Most prominent among these efforts was the initiative of a group of editors, including most of the major U.S. life science journals, who released a statement in February 2003. Noting the importance of protecting "the integrity of the scientific process by publishing manuscripts of high quality, in sufficient detail to permit reproducibility," as well as the "legitimate concerns about the potential abuse of published information" and the dual-use nature of much important research, the statement declared that *sometimes* self-censorship was appropriate:

> ". . . on occasion an editor may conclude that the potential harm of publication outweighs the potential societal benefits. Under such circumstances, the paper should be modified, or not be published" [30].

The statement was immediately criticised by both sides in the debate - those advocating great openness and those who would like to see more controls - which is an indication that it probably struck an appropriate balance.

Meanwhile, researchers themselves have become more conscious of the potential security implications of their work. After the American Society for Microbiology (ASM) changed the review policies for its eleven journals, requiring editors to be alert for sensitive information, they found that some authors began imposing self-censorship, for example, removing information on methods that might help others emulate the research. ASM's president, Ronald Atlas, argues that such actions, including proposals that editors limit or remove methods sections of research papers, can go too far: "Science, by its definition, is supposed to be repeatable, and if we permit publication of manuscripts that lack sufficient detail . . . we will be undercutting science," he declared in a newspaper interview [31]. Some scientists make the case even more strongly, arguing that instead of endangering national security, openness will improve security if "researchers have access to information that may lead to new vaccines, detectors, and treatments" [26]. The debate is far from settled and, like other aspects of the science-security balance, is likely to continue in years to come.

6. "Sensitive But Unclassified" Information

6.1 Government Policy on Classification of Information

In the federal government, the authority to limit the circulation of sensitive information in order to protect national security generally rests with the Executive Branch. Over the years, policy on classification of information has been defined through a series of executive orders issued by the President. Standards for classifying and declassifying information have tended to shift from one administration to another as the international situation and the political predispositions of the administration have changed. The current standards were defined by Executive Order 12,958, as amended in March 2003 [32].

Under these standards, information is classified as "Top Secret," "Secret," or "Confidential." The officer who classifies the information must justify his or her action based on the danger presented by its potential disclosure and must set a date after which the information must be declassified. Access to classified information is restricted to those who have received a security clearance and who have demonstrated a "need to know" the information. Federal law prescribes criminal penalties including prison sentences and fines for removing classified information without authorisation and for transmitting it to unauthorised persons.

The virtue of this dichotomous system of classification is its clarity - information is either classified or unclassified. There are definitions and procedures for handling it, and established penalties for violations. From time to time, however, the notion of an intermediate category known as "sensitive but unclassified" has arisen. Scientists in particular are resistant to this grey area, since they feel it leaves too much discretion to federal bureaucrats and runs counter to the ethic of openness in academic research.

6.2 Broadening the Concept of Classification

In November 2001, in response to requests from scientific leaders for assurances that Bush Administration was committed to openness despite the terrorist attacks and the new security environment (including the passage of the USA PATRIOT Act), National Security Advisor Condoleezza Rice cited the importance of "open and collaborative basic" scientific research in combating terrorism and reaffirmed that "the policy set forth in NSDD-189 shall remain in effect" in the U.S. government [33]. National Security Decision Directive 189 (NSDD-189), signed by former President Ronald Reagan in 1985, guaranteed that there would be "no restrictions… upon the conduct or reporting of federally-funded fundamental research that has not received national security classification" [34]. In other words, results of federally-funded research would continue to be either classified (and thus restricted in circulation) or unclassified (and available to all).

Shortly after Rice's letter, however, in March 2002, Andrew H. Card, Jr., chief of staff to President Bush, issued a memorandum to government agencies on safeguarding "sensitive but unclassified" (SBU) information, also known as "Sensitive Homeland Security Information" [35]. The "SBU" term later appeared in the Homeland Security Act of 2002 (Public Law No. 107-296) in Sections 891 and 892, which details procedures and guidelines for the sharing of such information. Consisting of unclassified but "sensitive" information, the SBU label serves as a means of "preserv[ing] confidentiality without formal classification" [35]. This is exactly what concerns many in the scientific community.

The federal government also has taken a number of other actions aimed at limiting the availability of information. In January 2002, for example, the Bush administration began withdrawing more than 6,500 declassified documents relating to sensitive chemical and biological warfare information from public access. In September 2002, after months of negotiations, the National Research Council (NRC) was finally allowed to publish a study it had conducted on agricultural bioterrorism for the U.S. Department of Agriculture (USDA). The NRC agreed to remove eight case studies from its report in deference to USDA concerns that the case studies could compromise national security by revealing vulnerabilities in the U.S. food supply system [36].

In March 2002, the Department of Defense circulated Security Directive 106 (SD106) on "Research and Technology Protection within the Department of Defense" detailing methods to safeguard Critical Research Technology and Critical Program Information at Department of Defense Research, Development, Test and Evaluation (RDT&E) sites [37]. The directive was retracted in May 2002 after objections by scientists at universities and other research institutions receiving DOD research funds to measures potentially restricting foreign travel, foreign collaboration, and publication of findings.

Federal agencies have removed documents from their websites, ordered that information at public libraries be removed from public access or destroyed, and, according to press reports, "stopped providing… information that used to be routinely released to the public" [38]. Information regarding the location of hazardous materials and facilities has been removed from public access, but "agencies have not articulated. . . the standards they used in choosing what information to make secret" [39]. Provisions in the Homeland Security Act of 2002 expand exemptions to public access to government information under the Freedom of Information Act by restricting disclosure of critical infrastructure vulnerabilities provided by private companies to the Department of Homeland Security (DHS). The provisions also protect disclosing companies from civil law suits and provide for criminal punishment of whistleblowers within government agencies who disclose company data to the public.

6.3 Continuing Controversy

The controversy over "sensitive but unclassified" information continues, not just in science but in many other sectors. A variety of interest groups and coalitions have spoken out in opposition to the concept. In August 2003, Rick Blum from the public interest group, OMB (Office of Management and Budget) Watch, wrote a letter on behalf of over 70 professional societies, civil rights organisations, special interest groups, and academic associations urging Secretary of Homeland Security Tom Ridge to allow public input into the process of defining "homeland security information" in accordance with the Homeland Security Information Sharing Act (HSISA) (P.L. 107-296) [40]. A study by the RAND Corporation released in May 2004 concluded that the federal government's efforts to remove information from its websites were ill-advised and ineffective. The U.S. government shut down 36 sites and over 600 public databases. RAND found that much of the information was not critical for terrorists, and could be obtained elsewhere, for example in textbooks, on non-government websites, in trade journals, or on maps [41].

Despite these objections and others, the federal government has expanded the SBU category. In June 2004, the Department of Homeland Security issued a draft management directive that would exempt federal agencies from releasing Environmental Impact Statements to the public as required by the National Environmental Policy Act (NEPA). Under the directive DHS could withhold the statements from the public if they contain sensitive information. However, the means by which DHS would identify such sensitive material in the reports are not clear [42].

H.R. 3550, a highway spending bill passed by the House in June 2004, contains a provision that would allow the government to keep secret records related to transportation of hazardous materials through cities [43]. The bill would supersede open-records laws at the state level, and give the Transportation Security Administration (TSA) wide latitude in defining what information is sensitive and should not be released. It would authorise the agency's director to withhold information deemed "detrimental to the safety of passengers in transportation, transportation facilities or infrastructure or transportation employees". Openness advocates have expressed concerns that the law would essentially suspend oversight of the TSA [44]. The bill did not receive final passage in 2004 but it is slated to be taken up again by Congress in 2005.

For its part, having learned to live with the system of classification and security clearance during the World War II and the Cold War, the science community is apprehensive that the introduction of a middle ground - "sensitive but unclassified" - will open the door to confusion, mistrust, and possible abuses by over-zealous officials. The unintended result of this trend, like that of other well-intentioned post-September 11 security initiatives, might be the opposite of what is expected. Rather than enhancing the nation's security, these initiatives could handicap American science and technology, retard its progress, and, in the end, diminish the nation's security - not just its homeland or military security - but its economic, environmental, and health security, and the quality of life of its citizens.

Notes

1 Based on research conducted in collaboration with Mark S. Frankel and assistance from Allison Chamberlain and Ryan Ricks. This material is based upon work supported by the National Science Foundation under Grant No. SES-0327825.

2 A select "biological agent" is defined by government regulation as "any microorganism..., or infectious substance, or any naturally occurring, bioengineered, or synthesized component of any such microorganism or infectious substance, capable of causing death, disease, or other biological malfunction in a human, animal, a plant, or another living organism; deterioration of food, water, equipment, supplies, or material of any kind; or deleterious alteration of the environment [9].

3 "Restricted persons" under the USA PATRIOT Act include individuals indicted or convicted of crimes punishable by imprisonment for more than one year, fugitives from justice and illegal aliens,

dishonorably discharged service members, any "unlawful user of any controlled substance," anyone who has been "adjudicated as a mental defective" or committed to any mental institution, or any national of a country deemed by the Secretary of State to support terrorism.

4 This section is based on an unpublished manuscript written in collaboration with Julie Fisher and Mark S. Frankel.

References

[1] American Council on Education, Association of American Universities, Council of Graduate Schools, Association of International Educators, National Association of State Universities and Land Grant Colleges. Survey of Applications by Prospective International Students to U.S. Higher Education Institutions; 2004. Available at http://tinyurl.com/68kqs.

[2] Brown HA, Syverson, PD. Findings from U.S. Graduate Schools on International Graduate Student Admissions Trends. Council of Graduate Schools. Summer 2004.

[3] US General Accounting Office. Border Security: Improvements Needed to Reduce Time Taken to Adjudicate Visas for Science Students and Scholars. Report. GAO-04-371. February 25, 2004. Available at http://www.gao.gov/new.items/d04371.pdf.

[4] Bollag B. Australia sees strong gains in enrollment of foreign students. The Chronicle of Higher Education. March 9, 2004.

[5] Payne D. Students blocked from US meeting. The Scientist. February 2, 2004.

[6] Szustaczek C. U.S. border laws keep professor home. University of Toronto, News@UofT. November 22, 2002. Available at http://www.news.utoronto.ca/bin3/021122b.asp.

[7] Statement and Recommendations on Visa Problems Harming America's Scientific, Economic and Security Interests. May 12, 2004. [Signed by 25 scientific, engineering, and higher education associations] http://www.aaas.org/news/releases/2004/0512visa.pdf.

[8] Harty M. (Assistant Secretary of State for Consular Affairs). We don't want to lose even one student. The Chronicle of Higher Education. Oct. 8, 2004. B-10.

[9] U.S. Department of Health and Human Services. Interim Final Rule on Possession, Use, and Transfer of Select Agents and Toxins. Federal Register. 240: 67. December 13, 2002, pp. 76886-76905.

[10] Wilkie D. Scientists turn from bioterror research. The San Diego Union-Tribune. June 1, 2004. Available at http://tinyurl.com/5yyqn.

[11] Gaudioso J, Salerno RM. Biosecurity and research: minimizing adverse impacts. Science. 304. April 30, 2004. 687.

[12] Borrego AM. Regulatory overkill? The Chronicle of Higher Education. January 31, 2003.

[13] Malakoff D. One year after: tighter security reshapes research. Science. 297. September 6, 2002. 1630-1633.

[14] Malakoff D, Enserink,M. Scientist on trial: Butler declared guilty on 47 of 69 counts . . . innocent of lying to the FBI and most smuggling charges. ScienceNow [online]. December 1, 2004. Available at http://sciencenow.sciencemag.org/feature/data/butlertrial.shtml

[15] Miller JD. Butler appeal ongoing. The Scientist: Daily News. December 22, 2004. Available at http://www.biomedcentral.com/news/20041221/02/

[16] Malakoff D. Student charged with possessing anthrax. Science. 297. August 2, 2002. 751-752.

[17] Miller JD, McCook, A. Artist bacteria case becoming costly. The Scientist: Daily News. December 21, 2004. Available at http://www.biomedcentral.com/news/20041222/03/.

[18] Committee on Research Standards and Practices to Prevent the Destructive Application of Biotechnology, Development, Security, and Cooperation. Biotechnology Research in an Age of Terrorism. National Research Council. Washington, DC: National Academies Press; 2004.

[19] US Department of Health and Human Services. HHS will lead government-wide effort to enhance biosecurity in 'dual use' research. (press release). March 4, 2004. Available at http://www.nih.gov/news/pr/mar2004/hhs-04.htm.

[20] Norris JT. Restrictions on Research Awards: Troublesome Clauses. Report of the Association of American Universities and the Council on Governmental Relations. April 8, 2004. Available at http://www.aau.edu/research/Rpt4.8.04.pdf.

[21] Buchanan W. Post-9/11 researchers fear muzzle from U.S. San Francisco Chronicle. January 3, 2003.

[22] US Department of Defense. Office of the Inspector General. Export-Controlled Technology at Contractor, University and Federally Funded Research and Development Center Facilities. D-2004-061. Available at http://www.dodig.osd.mil/audit/reports/fy04/04-061.pdf.

[23] US Department of Commerce. Bureau of Industry and Security. Office of Inspections and Program Evaluations. Deemed Export Controls May Not Stop the Transfer of Sensitive Technology to Foreign Nationals in the U.S. IPE-16176. Available at http://tinyurl.com/4r74e.
[24] Council on Governmental Relations and the Association of American Universities. Summary of University Concerns with Commerce IG Report. Unpublished. Used with permission. July 22, 2004.
[25] Monastersky R. Publish and perish? The Chronicle of Higher Education. October 11, 2002. A16.
[26] Shea DA. Balancing Scientific Publication and National Security Concerns: Issues for Congress. Congressional Research Service. Report RL31695. February 2, 2004.
[27] Blumenfeld L. Dissertation could be security threat. Washington Post. July 8, 2003. A-1.
[28] Zilinskas RA, Tucker JB. Limiting the contribution of the open scientific literature to the biological weapons threat. Journal of Homeland Security. December 2002.
[29] Broad WJ. U.S. is tightening rules on keeping scientific secrets. The New York Times. 17 February 2002.
[30] Journal Editors and Authors Group. Statement on Scientific Publication and Security. Science. 299. February 21, 2003. 1149.
[31] Schmid RE. Some scientists worry published research may be used by terrorists. Associated Press. July 27, 2002.
[32] Brooks N. The Protection of Classified Information: The Legal Framework. Congressional Research Service. Report RS21900. August 5, 2004.
[33] Atlas RM. National security and the biological research community. Science. 298. October 25, 2002. 753-754.
[34] U.S. President. National Security Decision Directive-189. National policy on the transfer of scientific, technical and engineering information. September 21, 1985. Available at http://www.fas.org/irp/offdocs/nsdd/nsdd-189.htm.
[35] OMB tackles sensitive but unclassified information. Secrecy News. 85. September 3, 2002. Available at http://www.fas.org/sgp/news/secrecy/2002/09/090302.html.
[36] National Research Council. Committee on Biological Threats to Agricultural Plants and Animals. Countering Agricultural Bioterrorism. Washington, DC. National Academies Press. 2002.
[37] US Assistant Secretary of Defense for Command, Control, Communications, and Intelligence. SD-106, 25 March 2002. (later titled Mandatory procedures for research and technology Protection within the DOD", DOD 5200.39-R). Available at http://www.fas.org/sgp/news/2002/04/dod5200_39r_dr.html.
[38] Matthews W. OMB weighs information classification. Federal Computer Week. September 16, 2002.
[39] Guzy GS. Are we protecting secrets or removing safeguards? The Washington Post. November 24, 2002. B-1.
[40] Letter from Open the Government Coalition to Secretary of Homeland Security Tom Ridge. August 27, 2003. Available at http://www.ombwatch.org/rtk/SBUletter.pdf.
[41] Baker, JC. et al. Mapping the Risks: Assessing the Homeland Security Implications of Publicly Available Geospatial Information. Santa Monica, CA: Rand Corporation. Report MG-142-NGA. 2004.
[42] DHS seeks exemptions from public disclosure requirements. OMB Watch. 28 June 2004. Available at http://www.ombwatch.org/article/articleview/2240/1/229?TopicID=1.
[43] H.R. 3550 Transportation Equity Act: A Legacy for Users (Engrossed as Agreed to or Passed by House). Available through http://thomas.loc.gov.
[44] Bruggers J. Highway bill's secrecy rules spark public-safety debate. Courier-Journal (Louisville, KY). June 27, 2004. Available at http://www.courier-journal.com/localnews/2004/06/27ky/A1-secret0627-10522.html.

Science and Technology Policies for the Anti-Terrorism Era
Andrew D. James (Ed.)
IOS Press, 2006

The Individual and Collective Roles Scientists Can Play in Strengthening International Treaties

THE ROYAL SOCIETY[1]
6-9 Carlton House Terrace,
London SW1Y 5AG, United Kingdom

Abstract. It is essential to support international agreements, such as the Biological Weapons Convention, through the formation of international scientific advisory panels to keep up with the rapid pace of technological advance in the relevant sciences. The research community must exercise judgment in the publication of their work and raise awareness of the ethical and legal requirements related to their research. There should be a clear objective of moving towards an international consensus on adopting appropriate codes of good practice, particularly in relation to their role in combating the diversion of science advances into activities that pose a threat to global security and peace. The existing legal constraints relating specifically to biological weapons development both nationally and internationally should be examined and consideration given to what needs to be done to strengthen such laws and how they can be built in to an enforceable code of practice.

1. Introduction

A major challenge facing societies today is the threat to global peace and security which can be perpetrated by special interest groups or by nations acting illegally. Of particular concern is the potential use of sophisticated weapons based on cutting edge science in the fields of nuclear, chemical and biological technology. This discussion paper outlines the individual and collective roles scientists can play in strengthening international treaties aimed at preventing the proliferation and use of chemical, biological, radiological or nuclear weapons. There is a range of issues to be addressed, from the potential for science to be misused to its role in risk reduction and mitigation. There is a need for the scientific community, governments and relevant agencies to be fully aware of the potential of scientific advances both in enabling the illegal development of more lethal weapons and in developing more effective counter measures to the use of such weapons. Measures are needed to ensure that governments and relevant agencies have access to informed scientific advice. In addition, scientists need to be informed about the potential misuse of science and their responsibilities in meeting the requirements of international treaties and conventions aimed at preventing the proliferation and use of chemical, biological, radiological or nuclear weapons. The need to underpin the Biological Weapons Convention (BWC) is emphasised, because of the use in biomedical science of potentially harmful pathogens and toxins and the risks of this research being misused in bioterrorist attacks.

2. Scientific Underpinning of International Treaties and Conventions

2.1 The Role of Advisory Panels

In countries with a developed technological infrastructure basic science research is largely publicly funded through government agencies. In addition, major areas of technology development perceived to be in the national interest in both defence and civil sectors often receive substantial public funds. Thus governments have direct routes to the research community and often use this as a means of accessing science advice including via science advice panels. Even so, such means can be ad hoc with important sources of expertise being overlooked. On an international scale the situation can be even patchier, although there are examples of successful advisory bodies such as the International Panel on Climate Change, the European Pharmacopoeia Commission and various science advisory groups supporting the World Health Organisation.

Threats posed by the proliferation and use of chemical, biological, radiological or nuclear weapons can at the most sophisticated levels involve cutting edge science and technology. International treaties and conventions aimed at combating such threats must therefore incorporate as far as reasonably practical sound scientific principles. Access to the best science advice is essential and an important framework for achieving this can be an international science advisory panel with access through the membership to nationally based science expertise. Again there are examples where international agreements are backed by access to cutting edge science such as the Organisation for the Prohibition of Chemical Weapons (OPCW), which informs the Chemical Weapons Convention (CWC), and the International Atomic Energy Agency (IAEA), which sponsors research and development (R&D) to underpin decisions taken with the Nuclear Non-Proliferation Treaty (NPT).

The Biological Weapons Convention (BWC) has encountered significant problems in its effective implementation because of a lack of agreement on verification procedures to ensure that the Parties comply with the rules set out in the Convention. A particular issue is that laboratories and installations connected to the BWC are more diffuse and harder to monitor than those connected with for example nuclear materials. There is also the consideration that many agents may have "dual use" application, i.e. some research unconnected with biological and toxin weapons may also have a military value. Similar "dual use" issues might arise in the chemical and nuclear industries.

Meeting these challenges and developing measures to counter the use of biological weapons by terrorist groups and rogue states requires access to scientific knowledge at the forefront of biotechnology. However, currently there is no equivalent international organisation, such as the IAEA or the OPCW, supporting the BWC to ensure access to leading edge science. We believe that such support is essential and the framework for providing this is through the formation of an international advisory panel that is able to keep up with the rapid pace of technological advance in the life sciences.

2.2 Key Features of Successful International Science Advisory Panels

Such groups have in common highly respected memberships working under the authority of bodies set up with international political agreement. They also work to a clear set of objectives that are widely accepted as beneficial to human welfare. The key requirements can be summarised as follows.

- Expertise: The membership must have international status and represent a broad perspective on the science relevant to the area under consideration. Rotation of membership is beneficial, particularly in maintaining knowledge of scientific advances in the States Parties;

- Independence: Transparency on potential "conflicts of interest", for example from industrial affiliations, is essential;
- Personal Attributes: The group requires commitment, vision, and a strong motivation to engage in debate to arrive at a collective view;
- Strong Leadership: The chair of the group should have a clear understanding of both the intellectual and political issues, as well as an ability to motivate the group and to optimise its effectiveness;
- Mandate: A clear mission and terms of reference defining scope, aims and accountabilities are required;
- Size: A balance must be made between keeping the group to a manageable size and having a sufficiently broad representation to have international influence with Governments.

3. Compliance with Legal and Treaty Requirements

3.1 Scientists' Responsibilities

The scientific community must be aware of its responsibilities to work within ethical boundaries and comply with the requirements of both national legislation and international treaties and conventions. While this responsibility is clear in excluding the illegal development of weapons systems it is less clear in areas where there is the potential for dual use of scientific advances such as in biotechnology. The scientific community both at the individual and institutional levels must understand the requirements of national and international laws as they apply to its work. This is well established, at least in the technologically developed countries in the case of health and safety, with robust assurance and auditing functions in place even in small R&D laboratories. However there is often less awareness among the research community of the requirements of international treaties and conventions such as the BWC.

As a further measure to reduce the risk of illegal activities and key data and material getting into the wrong hands there is a need for a more systematic approach through education and communication to ensure that the scientific community is properly informed. In October 2004, the Royal Society held a joint meeting with the Wellcome Trust to identify measures that would help the wider life science community address concerns over biosecurity and bioterrorism, and contribute to the 2005 BWC Annual Meeting chaired by the UK Government.

3.2 Institutional Constraints

The enormous expansion in the life sciences coupled to the concerns about the potential for developing biological weapons capable of causing major societal damage and chaos has stimulated discussions on the need for more rigorous regulation to filter out research that could lead to such weapons. The key questions relating to institutional constraints are as follows:

- Should publicly and industrially sponsored research proposals be subject to an additional layer of vetting (in addition to the traditional evaluation of excellence and timeliness) with the objective of preventing potentially harmful research being carried out?

- Should papers submitted for publication, particularly in the life sciences be subject to a further layer of vetting to prevent diversion of the information into harmful applications?

Such filtering is clearly appropriate in the case of research proposals and papers where there is a tangible cause for concern in terms of harmful applications. However, this is probably best achieved on a case by case basis by the relevant sponsors and journal editors. Going beyond this, for example in applying a vetting process across the spectrum of basic research proposals, even where there may be some, although unidentified, dual use potential, would be difficult and impose a burdensome layer of bureaucracy on the research enterprise. For example, how would such a process have applied to the fundamental nuclear physics research proposals in the 1920s and 1930s that provided the foundations for the development of nuclear weapons? Equally difficult is the filtering of basic research papers on grounds of a potential threat to security. This was highlighted by the *Statement on Scientific Publication and Security* published by the editors of a number of leading scientific journals including *Science*, *Nature* and the *Proceedings of the National Academy of Sciences*. They advocate increased vigilance in identifying papers where the potential harm of publication outweighs the potential societal benefits but they do not identify hard and fast rules for doing this. The choice of actions will depend on the judgement of the editors and their referees. Nevertheless the research community must exercise judgement in the publication of their work and this emphasises the need to raise the awareness of the science community of ethical and legal requirements related to their work.

4. Codes of Conduct

This is a contentious issue that is receiving increasing attention, stimulated by the potential for dual use, particularly in the life sciences. As with institutional constraints there is the danger of a substantial and complex bureaucratic process being imposed on science that would be extremely difficult to manage and would have limited value in reducing the risk of science being misused.

4.1 General Considerations

A clear and agreed understanding is needed of what is meant by codes whether of ethics, conduct or practice. A code in engineering means a rigorous set of rules governing compliance with such things as design, construction and operating practice. Failure to do so can result in legal action against the relevant bodies, i.e. design authority, constructor or operator, particularly if safety has been breached or there is a major loss of investment. Thus an essential question for codes governing scientific practice is that of enforcement.

Researchers in the UK and elsewhere must comply with safety rules and ethical standards as specified by ethics committees relating for example to experiments involving humans and animals. Rules also apply to issues of integrity as in the honest recording of results, plagiarism and declaration of relevant interests in published papers. Breaching the rules is subject to sanction and there have been some high profile cases particularly in the misrepresentation of results and failure to meet ethical standards in published papers. However, the means of detecting such breaches have been rather ad hoc and this highlights the difficulty of having a rigorous process of vetting and enforcement. Nevertheless there is a case for examining the need for having more rigorous codes applied to scientific practice and its role in potentially reducing the risk of misuse for illegal weapons development.

More specifically, the essential elements of good practice can be defined and incorporated into more rigorous codes governing the execution and reporting of scientific research. Such codes while containing many common elements will also have elements appropriate to the field in question e.g. bioscience, chemistry and nuclear physics. The common elements could encompass meeting general safety and ethical standards such as potential conflicts of interests, plagiarism and misrepresenting or exercising bias in recording and publishing data, as well as practical requirements such as the keeping of comprehensive and auditable laboratory records. Specific elements may cover aspects of safety and security such as the handling of potentially dangerous materials. Good practice should also include the responsibility of scientists to be aware of and comply with the requirements of international conventions and treaties in their research area. This needs educational and research institutions to put in place the appropriate measures to enable this requirement to be met.

Going beyond this to define and apply enforceable codes governing wider ethical and moral aspects related to good conduct would be extremely difficult. Nevertheless, there would be merit in giving careful consideration to identifying guiding principles that should be used by researchers in the conduct of their research. These would cover areas relating to personal integrity beyond those referred to above in meeting their responsibilities to society, for example in carefully identifying and communicating the balance of risks associated with research outputs. Clearly this should include the potential misuse of their results for illegal weapons development, in so far as such potential is discernible when the research is done.

4.2 Application of Codes and Guiding Principles

Careful consideration needs to be given to whether a code of good practice will be effective. This includes questions such as how the code and good practice procedures will be enforced, who will be responsible for checking a researcher's work, what penalties would occur if a researcher contravened the code, whether "whistle blowing" would be encouraged, and what mechanisms would be in place to protect the whistleblower. There are a number of examples of codes of conduct in fields of science that could be used as a model, perhaps the best known in the UK being the General Medical Council's code of ethics for doctors. Many professional organisations have required members to subscribe to a code of conduct for a number of years (e.g. UK Institute of Electrical Engineers since 1972, the American Society of Microbiology since 1988, American Chemical Society since 1965). These codes include consideration of the member's role in serving society's interest. Guidance on professional practice is also commonly available, for example for microbiologists to keep written records of all requests for reagents, technologies and knowledge, and to monitor such requests and derive a risk assessment before deciding whether or not to fulfil a request. This raises the question of checking how such procedures would be upheld: a practical answer is for individual research institutes to be responsible for application of the rules in the execution of research. Only when the work is exposed to the external world through publication and/or application is there a need for a wider examination of any breaches of codes of practice.

4.3 Codes and International Agreements

There should be a clear objective of moving towards an international consensus on adopting appropriate codes of good practice, particularly in relation to their role in combating the diversion of science advances into activities that pose a threat to global security and peace. This is a formidable task but one way forward would be for international agreements aimed at preventing illegal weapons development and application to be underpinned through the incorporation of such codes with each government acting as guarantor. Given its present state of development there is an opportunity to take this forward in the BWC alongside the proposal to set up an international advisory panel. It would require the States Parties to work towards defining an agreed code and to demonstrate their commitment through the setting up of the processes to ensure compliance. In moving towards this objective it would be worth examining the existing legal constraints relating specifically to biological weapons development both nationally and internationally and considering what needs to be done to strengthen such laws and how they can be built in to an enforceable code of practice.[2]

Notes

1 This paper was originally produced for the United Nations Foundation, Nuclear Threat Initiative (NTI) and National Academies peer review round table on biological threats to security, held in Washington DC, on 19 April 2004 (see www.un-globalsecurity.org for more details), and represents the views of the Royal Society, the UK national academy of science.

2 The Royal Society has produced the following policy reports and submissions relating to the scientific aspects of international security. All of these documents are available online at www.royalsoc.ac.uk

- The Roles of Codes of Conduct in Preventing the Misuse of Scientific Research; June 2005.
- Issues for Discussion at the 2005 Meeting of Experts of the Biological and Toxin Weapons Convention; June 2005.
- Do No Harm: Reducing the Potential for the Misuse of Life Science Research. Report of a Royal Society – Wellcome Trust meeting held at the Royal Society on 7 October 2004; December 2004.
- Making the UK Safer: Detecting and Decontaminating Chemical and Biological Agents; April 2004.
- Response to the House of Lords Science & Technology Committee Inquiry into Science and International Agreements; January 2004.
- Response to the House of Commons Science & Technology Select Committee Inquiry into the Scientific Response to Terrorism; February 2003.
- Response to UK Foreign & Commonwealth Office Green Paper on Strengthening the Biological Weapons Convention; November 2002.
- Joint statement from the Presidents of the US National Academy of Sciences & the Royal Society, Bruce Alberts and Lord May. Scientist support for biological weapons controls; November 2002.
- The Health Hazards of Depleted Uranium Munitions Part II; March 2002.
- Royal Society Foreign Secretary Sir Brian Heap's editorial, Scientists against biological weapons. Science 2001; 294: 1417.
- The Health Hazards of Depleted Uranium Munitions Part I; May 2001.
- Measures for Controlling the Threat from Biological Weapons; July 2000.
- Management of Separated Plutonium; February 1998.
- Scientific Aspects of Control of Biological Weapons; July 1994.

Science and Technology Policies for the Anti-Terrorism Era
Andrew D. James (Ed.)
IOS Press, 2006

National Security, Terrorism and the Control of Life Science Research

Brian RAPPERT
Department of Sociology, University of Exeter,
Exeter EX4 4QJ, United Kingdom

Abstract. This Chapter assesses ongoing attempts to balance security and openness in the conduct of civilian bioscience and biomedical research. More specifically it examines the state of current policy discussions regarding the security threats posed by life science research results and techniques. Concern about their "dual-use" potential has intensified tremendously since 9/11 and the anthrax attacks in the U.S. Despite the considerable attention being paid to this issue today, this Chapter argues that current discussions are unclear and arguably problematic in relation to vital questions regarding the problem posed by life science research, the ultimate goals of controls, and the desirability of the circulation of dual-use knowledge. It seeks to challenge the state of discussion by asking how the current security presentations of the threats posed by emerging bioscience developments are themselves formed in relation to definitions of scientific practice. On the basis of this analysis, consideration is given to the potential and pitfalls associated with current international efforts to devise a "code of conduct" for bioscientists.

1. Introduction

Today in many countries, renewed attention is being given to the potential for the biological and medical sciences to facilitate the development of biological weapons.[1] Questions are being asked as to whether some research is too "contentious" to pursue and what systems should be in place to evaluate and possibly restrict laboratory activities. With this has come a questioning of whether the scrutiny and controls being undertaken are damaging the science base. Debates about what, if any, controls might be prudent often turn on how classifications and distinctions are made in relation to questions such as: What counts as "dangerous" research? In what ways is the understanding of life processes now being generated opening up new ways of interfering with the functioning of humans, animals, and plants? What research findings constitute a novel contribution to a discipline? To what extent are the security issues associated with the life sciences similar or different to those associated with other Weapons of Mass Destruction (WMD) areas of concern such as nuclear physics?

This Chapter considers what is at stake in the way such questions are being posed and answered today. It does so by questioning how "the problem" with bioresearch is defined. In particular it asks how current depictions of the threats posed by emerging bioscience developments are themselves formed in relation to understandings of scientific practice. It aims to identify key issues hitherto not given sufficient attention, the knots and binds associated with trying to assess the dual-use capacity of research, and problems for future consideration. Following from an analysis of these issues, the pros and cons associated with attempts to develop an international code of conduct for bioscientists are discussed.

This Chapter itself employs several distinctions to set the boundaries of analysis. The first is between research controls intended to prevent the unauthorised acquisition of biological agents or their accidental release with those controls intended to prevent the knowledge and techniques generated from being misused. While the first refers to the fairly well trodden ground of biosecurity/biosafety, the second raises much less familiar matters for the life sciences. It is the latter which is the focus here. In addition, this Chapter is concerned with the implications of controls for non-military agency funded academic (and to a lesser extent) industrial R&D [1]. Finally, much of the argument centres on recent activities in the U.S., the country which is at the forefront of debating and adopting research controls. At the time of writing while European countries had made various regulatory biosecurity/biosafety reforms, discussions are only beginning about "dangerous knowledge". This Chapter's focus on the U.S. is appropriate given that it is the site of a considerable percentage of the world's life science research and because of its ongoing interest in internationalising any controls adopted in the U.S.

2. The Life Sciences and Biological Weapons

Discussions about the dual-use potential of biological and medical research today include concerns that such activities may facilitate state and non-state actors making bacteria more resistant to antibiotics, modifying agents' virulence and pathogenicity, synthesising viruses, devising novel bioweapons, and reducing the effectiveness of the body's defence system [2]. By dual-use is meant their potential for civilian and military applications and while renewed attention is being paid to the proliferation of nuclear and chemical expertise and equipment, there is little doubt that much of current public policy attention regarding the security implications of sciences centres on the life sciences.

2.1 The Distinctiveness of Life Science Research

The impetus behind and character of this focus derives from a series of distinctions made between life science research and other "dual-use" areas of concern such as nuclear science and cryptography. In recently assessing the threat posed by biological weapons used by terrorists, for instance, the U.S. National Research Council (NRC) concluded that the dual-use potential of life science research is much more of a potential problem than that posed by nuclear physics [3, p.112]. This is because of comparative differences in the steps and resources required between fundamental science and the production of a weapon, including: the accessibility of the materials required, the ease of distinguishing civilian from military relevant activities, the size and global dispersion of the communities, and the extent of the incorporation of security issues within the existing culture of research [4]. In brief, the NRC argued that the resources and knowledge required for the production of biological weapons is more thoroughly and troublingly "dual-use" than that necessary for nuclear weapons [5]. Much the same overall conclusion was said to apply to cryptography. As another consideration, the diffuse nature of the materials and expertise in the life sciences were said to limit the viability of research controls.

The importance of these distinctions for security responses depends on the nature of the would-be user of the weapons. While much attention has been given to terrorist threats in recent years, this did not figure as a prominent topic in the past.[2] However, the aforementioned differences in the dual-use potential of research would suggest that the materials and expertise necessary for attacks with biological weapons are far more readily accessible than those required for devising nuclear weapons. For some though, the concentration on terrorism today misplaces the most likely developers of bioweapons: namely states in offensive programmes. Among the demands often cited for producing a viable bioweapon include obtaining a virulent strain of an agent, culturing it in sufficient quantities, weaponising the agent (e.g. through aerosolisation techniques) and then securing the means of dispersing it. Many security analysts have taken the limited employment of biological weapons around the world despite the historical interest in such capacities as an indication of the difficulty of effectively employing them [7, 8]. In sum, there is much debate today about the characteristics of life science research and the likelihood of the threats posed.

2.2 Policy Responses

Much of the initial policy response post 9/11 and the anthrax attacks in the U.S. has been directed towards strengthening the physical containment of pathogens. For instance, the 2001 U.S. PATRIOT Act, the U.S. Public Health Security and Bioterrorism Preparedness and Response Act of 2002 and the 2001 UK Anti-terrorism, Crime and Security Act brought in enhanced controls on who could access traditional dangerous pathogens and toxins (mainly so-called "select agents" by U.S. Center for Disease Control classification).

However, in the U.S. at least, the debate has expanded significantly beyond issues associated with personnel and materials to consider how the knowledge and techniques generated from research might pose security concerns. What is often said to be the relative ease of moving from the science to the production of biological weapons as well as the pace of scientific developments has led to questions about the advisability of pursuing some lines of research. Post 9-11, experiments such as the insertion of the interleukin-4 gene into the mousepox virus, the comparison of *variola major* and *vaccinia* viruses and the artificial chemical synthesis of the polio virus have routinely figured as examples of how non-applied research might facilitate the development of biological weapons [9, 10]. With the passage of the Homeland Security Act in the U.S., renewed attention has been given to the possibility of introducing a "sensitive but unclassified" security categorisation for certain research findings [11]. In early 2003, a group of 32 largely American based scientific journals met to agree guidelines for reviewing, modifying, and perhaps even rejecting research articles where "the potential harm of publication outweighs the potential societal benefits" [12]. In 2004, the NRC's report *Biotechnology Research in an Age of Terrorism* made significant new recommendations for the review and oversight of potentially dangerous research proposals. While it found existing legislation and regulation for protection of materials and the vetting of personnel adequate, it concluded that additional procedures should be in place to screen so-called "Experiments of Concern" [3, p.112]. Currently this category includes activities such as increasing the transmissibility of pathogens, enhancing virulence of agents, and rendering vaccines ineffective; although the Council has stated that the classification might have to expand in the future. The National Research Council suggested this category of experiments be reviewed through expanding the remit and expertise of existing local Institutional Biosafety Committees (IBCs) in the first instance, with the second stage referral of problematic dual-use cases to an expanded Recombinant DNA Advisory Committee. The U.S. National Science Advisory Board for Biosecurity (NSABB) was set up in 2004, in large part to take forward the recommendations in *Biotechnology Research in an Age of Terrorism.* While most of the initiatives involving controls on dual-use research findings have taken place in the U.S., there has been active attention to whether and how these might be internationalised in forums such as the OECD International Futures Programme and the International Forum on Biosecurity.

3. Issues for the Control of "Dual-Use" Knowledge

Thus, in addition to a tightening of the procedures for recording and approving who works with what materials in which labs, questions are being asked about the advisability of conducting and communicating what (in another security environment) might be regarded as unremarkable research. Arguably, the situation today is characterised as one of unprecedented attention in the life sciences on how to integrate security and openness. The long held assumption in relation to civilian scientific research vis-à-vis biological weapons – that national security is best served by the beneficial and protective innovations deriving from unfettered research – is increasingly being questioned as the biosciences are told to "lose their innocence" [13]. Despite the significant amount of attention being paid to these issues today, this Chapter argues that the discussion to date has been characterised by ambiguity regarding its central premises and objectives. In the following section, it will identify some of the problematic issues associated with the "silencing of science" as currently proposed [14].

3.1 What is the Goal?

Although some members of the bioscience community have criticised the proposed oversight measures as imposing unnecessary and counter productive burdens, it is not clear how or even if the system being discussed in the U.S. today will curtail avenues of investigation [15]. While the 2004 NRC report calls for existing IBCs to factor security concerns into their decision-making processes, it was decidedly vague about just what reviewers should be looking for in this process. The NRC calls for "carefully weighing the potential benefits versus the potential danger" of research proposals [3, p.91] without any further detail as to how experiments should be assessed. What such a general call should mean in practice is far from clear though. A similar lack of specifics has characterised other oversight proposals. For instance, the Center for International and Security Studies at the University of Maryland proposed the idea of a "Biological Research Security System" in which local approval committees were simply tasked with considering "the scientific merit of the proposal, the experience of the principal investigator, and the potential risks" [8]. However, such steps are already taken to assess the biosafety of experiments. The task of specifying rules for assessment is a daunting one given the complexities of the issues at stake in evaluating the dual-use potential of research, The U.S. NSABB has been charged with developing guidelines, though at the time of writing this process was just beginning. Major criticisms have also been made of the frequency, rigour, and transparency of U.S. IBCs [16].

Despite the present uncertainty about just what the review process will entail, there is good reason for believing that any system which takes as its basic decision logic the weighing of risks versus benefits is unlikely to impose many or nearly any restrictions. A major reason for this is the importance attributed to staying ahead of security threats through rapid innovation in detection, prevention and treatment measures. Following this line of thinking, in order to prepare for bioattacks by terrorists or others, deliberate steps have to be taken to understand the nature of the threats faced. Each of the three contentious experiments mentioned previously (artificial polio virus, IL-4 and mousepox, and the comparison of *variola major* and *vaccinia* viruses) were strongly defended by the scientists involved and others as providing much needed information about disease mechanisms that could (if acted upon with yet further R&D) alleviate or prevent the consequences of attacks [17, 18, 19]. If what is helpful for terrorists can also be so for counter-terrorists, then in practice when it comes to making determinations of who can make the best use of research, it is almost certainly going to be decided that this rests with the counter-terrorists because we can call on disproportionate expertise, resources, and abilities. So even in the case of publishing research on how to make a ciproflaxin[3]-resistant *bacillus anthracis* the editor of the *Proceedings of the National Academy of Science* has suggested that such work may well be important to publish because of what it could reveal about the characteristics of resistant strains [20].

The importance attributed to what might be characterised as the "run faster" strategy is amply illustrated by considering the recent U.S. emphasis on biodefence R&D. While in FY2001 U.S. civilian biodefence funding totalled $414 million, in FY2005 it is budgeted at over $7.5 billion [21]. Research on infectious disease is a major strand of the biodefence response. The National Institute of Allergy and Infectious Diseases at the National Institutes of Health is leading much of this work. Its budget has gone from $53 million in FY2001 to $1.7 billion in FY2005. This represents the largest ever increase in funding for a single initiative at the National Institutes of Health. The research is focused on "Category A" traditional agents (anthrax, smallpox, tularemia, plague and so forth) and broken down into research focused on therapeutics, diagnostics, host response, vaccines, basic biological mechanisms and building expert resources. Some have suggested that this increase in funding constitutes a disproportionate response given that it is unlikely that terrorists on their own could successfully deploy bioweapons [22]. Others have condemned it as far too limited arguing that the "revolution in the life sciences" will mean that the future threat may go well beyond those traditional agents and thus the response needs to be more comprehensive (see next sub-section) [23, 24]. In addition, other military and non-military agencies are funding their own work. The National Biodefense Analysis and Counter Measures Center under the Department of Homeland Security has been tasked to function as a "hub and spoke" agency for devising countermeasures. It is currently planning to conduct threat assessments that include analysing the storage and dispersion of biological agents as well as providing high-fidelity modelling and simulating disease transmission [25]. They signal an active interest in exploring varied threats from biological attack even leaving aside the questionable permissibility of such threat assessment activities under the Biological and Toxin Weapons Convention (BTWC).

While this massive increase in biodefence funding is consistent with the basic underlying logic of running faster, it is rather at odds with the assumption that the forthcoming oversight system will limit lines of research. Given the extensive activities underway and the overall legitimacy (indeed urgency) accorded to biodefence measures, it is far from clear in the future that IBCs will be asked to screen out research that might directly raise dual-use issues. Examining such topics is a central aim of extensive new streams of funding in the U.S.[4] Moreover, with the emphasis placed on the importance of innovation and "staying ahead" of possible threats where the dual-use concerns are most acute, the calls for unfettered research are likely to be most pronounced.

3.2 Where is the Problem?

Many of the existing or proposed controls on dual-use research rest on the ability to distinguish between problematic/unproblematic, acceptable/unacceptable, and potentially dangerous/benign knowledge. Those bits of information identified as particularly troubling are then to receive a further level of scrutiny and possibly restriction. While this approach has a certain straightforward logic, it is not clear that it adequately captures the issues at stake.

One problem is the difficulty of identifying contentious research against the backdrop of other work taking place. While in the abstract it is often commented that nearly all research poses dual-use concerns, efforts to identify actual significant instances of dual-use knowledge have proved more elusive. Journals such as the *Proceedings of the National Academy of Sciences* (PNAS) and the eleven journals of the American Society for Microbiology (ASM) have had procedures in place providing additional scrutiny for dual-use manuscripts since 2002. However, relatively few manuscripts have been deemed to require closer scrutiny. Of the 22,486 manuscripts received by ASM journals between January 2002-July 2003, 575 manuscripts related to select agents were chosen for special screening, of which 2 were sent to the full ASM publication board (its second tier of scrutiny) for possible modification [3, p.71]. In a two month period during late 2002, the *PNAS* identified 1% of its manuscripts as requiring special attention by its Board, though no changes were made to any of these submissions [20, p.1463].

Even if the three controversial publications mentioned above had gone through the procedures agreed in 2003, it seems unlikely that any modifications would have been imposed. Part of the reason for this relates to the points noted in the last sub-section about how risks and benefits are likely to be assessed. Another reason relates to the manner in which individual publications are likely to be evaluated against others [26]. For instance, when the former editor of the *Journal of Virology* was asked to justify the publication of the IL-4 mousepox article, he argued that its findings should not be seen in isolation from what was already "out there" in the literature [19]. Rather, it had to be seen against prior work into IL-4 and pox viruses and an understanding of what someone with the requisite skills might be able to ascertain from the literature. When approached in this fashion, it became quite difficult to draw a line demarcating the "acceptable" from the "unacceptable". Halting the mousepox publication, the former editor argued, would have been ineffective because the experiment discussed in it built on previous work in the field and its results could have been guessed by those knowledgeable about the issues at hand. Yet, the further one moves away from the particular contentious results in question, the more one would have to restrict whole areas of research, in this case investigations into the basic mechanisms of immunology. Strikingly similar lines of reasoning, it can be argued, have been offered for other controversial cases. As such, it becomes quite problematic to justify isolating bits of research that are somehow in-and-of-themselves problematic. Questions can always be asked (and often with good reason) about what is already foreseeable and derivable.

One solution proposed by the former editor of the *Journal of Virology* was to focus on those areas of research that throw up unexpected findings. Yet this suggestion is problematic because what counts as a continuity or discontinuity can be and has been a matter of disagreement. In the case of the mousepox experiments, despite suggestions after the fact that the researchers should have been able to predict the results of their experiments [27], the Australian researchers are reported to have said the results were a surprise [28]. What one "skilled in the art" might "reasonably" be able to ascertain from the cumulative findings in the literature is not straightforward. Research in the field of Science and Technology Studies on claims to scientific novelty and replication would suggest that much interpretative work goes into determining what experimental results count as a continuity or discontinuity [29, 30].

Perhaps a more fundamental issue is the manner in which efforts to identify particularly troubling instances of dual-use research fail to address the totality of developments in the life sciences. Many of the existing research controls have been geared towards select agents. However, some commentators argue that traditional agents (or even traditional agents with modified virulence, antibiotic resistance or transmissibility profiles through recombinant DNA technology), represent only a fraction of possible future threats from biological weapons. As contended, the major problem for the medium term lies with the ability to cripple individuals' immune systems, develop "stealth" viral vectors with gene therapy technology, produce highly specific toxins with transgenic plants, fashion novel agents, and target the neurological system. These, they argue, are all options soon to be afforded by the "revolution" or "paradigm shift" in the life sciences. Each stage of the use of bioweapons – their production, weaponisation and delivery – will be eased through future research in the life sciences as well as a host of related fields [2, 31, 32]. With the proliferation of skills and the simplification of biotechnology techniques comes the possibility that a wider range of individuals from terrorists to sociopaths will have the ability to cause mass disruption if not death [33].

Thinking in terms of such large scale and systematic changes though requires moving well beyond a focus on particular experiments and traditional select agents. The major source of concern is not the odd experiment; rather it is with the cumulative "advance" of academic and industrial R&D and what this means for the development of enabling technologies and the rapid diffusion of capabilities. If the next generation threats being discussed today are realistic options for skilled terrorists within a medium time frame, then seeking to identify (an already quite limited range of) "contentious" experiments that might be suggestive of dual-use possibilities is deeply limited and missing the point.

3.3 How should Dual-Use Concerns be Communicated?

The last two sections have emphasised the intense efforts underway in biodefence and the potentially widespread concerns associated with life science research. This raises questions about how the implications of research should be communicated to both specialised expert audiences and the public in general.

For those critical of attempts to impose "non-proliferation-type" controls and in favour of the "run faster" response, limitations on performing or publishing research because of dual-use fears will ultimately prove to be counterproductive. The extensive biodefence programmes emerging in the U.S. are predicated on the wisdom of attempting to stay ahead of potential threats through rapid scientific and technological advancement. With the breadth of activities budgeted for – from fundamental research on the functioning of cells to highly applied research on threat assessments about the aerosolisation of agents – the next few years are likely to see the production of many significant "dual-use" findings. How those should be communicated remains a topic of uncertainty and contention. If significant elements of U.S. biodefence activities are kept secret (particularly those close to weaponisation) then this is likely to arouse suspicion in other countries regarding the ultimate aims of the U.S. programmes. In the past, heated claims and counterclaims about the legitimacy and purpose of "biodefence" programmes have figured as part of international disagreements about whether countries are adhering to the BTWC. Without some level of transparency, the sheer scale of activities now taking place in the U.S. will almost certainly cause suspicion in other countries and give them reason to reconsider whether their biodefence measures are adequate or even too transparent. Yet, if the results of threat assessments or other such studies directly related to the functioning of biological weapons are made more widely available to encourage the development of responsive measures (as would be consistent with the "run faster" approach) then such information could be quite valuable to would-be bioweapons engineers in both developing devices and refining them to evade detection or treatment measures.

On a more subtle level, the question is not just whether research should be published or not but whether as part of that process possible dual-use implications should be elaborated. Nearly all of the fundamental research conducted by the NIH is likely to be published and – if the argument put forward in this Chapter is accepted – that research is likely to be published without modification. However, whether the possible dual-use implications of those experiments and studies should be actively discussed is another matter. Such a practice, though consistent with the goal of raising general awareness in the life sciences and with the importance of community-wide responses, would almost certainly lead to widespread public concern about the direction and merit of life science research. Experiments where some level of accessible speculation has taken place regarding the implications of the findings, such as the experiments with artificial polio virus, IL-4 in mousepox, and the comparison of *variola major* and *vaccinia* viruses have already generated calls by politicians and others for controls deemed draconian and ill-advised by some of those in the scientific community [34]. Should bioscientists routinely hypothesise about the possible malign utility of their work, (unwanted) critical public and political attention will surely increase. For instance, the artificial polio virus publication in 2002 has been highly criticised as an opportunistic PR stunt because it was said to have been known for years if not decades that this sort of synthesis was possible [35]. Whatever the merits of such denunciations, arguably few outside of life science research knew of such capabilities or considered their implications. Should concerted efforts be made by scientists in forthcoming years to elaborate possible dual-use implications of their work (even if the research in question is not "new"), then a considerable reaction is likely.

3.4 Who Decides?

Throughout the discussion above, questions have been raised regarding who should make decisions about how to balance research and national security. Despite frequent complaints about governments imposing controls on research, when it comes to matters involving restrictions on knowledge, even in the U.S., government agencies have been reluctant to compel bioscientists to implement security controls, let alone stop lines of investigation. There are few, if any, bright lines indicating what research is proscribed. Instead, the scientific communities have been asked to formulate ideas for what needs doing. As suggested above though, the issues at stake are not simply technical matters that can be resolved through the application of expert knowledge. Differences exist in what one "skilled in the art" might reasonably know, what counts as a realistic future weapons possibility, and what research should be permissible. So while scientific expertise is obviously essential in evaluating threats and considering policy options, questions can be raised as to whether it is sufficient. The National Research Council recommendations now being taken forward in the U.S. conceive of oversight decision as the prerogative of scientists. In contrast, the Maryland Center for International and Security Studies proposal for a "Biological Research Security System" calls for "a mix of scientists and public representatives who are not directly involved in the research in question" [8].

4. What Prospect for "Codes of Conduct"?

Against the backdrop given above, this section examines the prospects associated with developing a "code of conduct" for bioscientists and others associated with the life sciences. While the idea of formulating a professional ethical code has been mooted for decades [35], since 2001 various statements by governments, NGOs and professional societies have given renewed impetus to this idea.[5] The adoption of some sort of code has become incorporated into the formal agenda of key institutions in and outside the U.S. In 2005, as part of the talks being held under the Biological and Toxin Weapons Convention, the topic for discussion was "The content, promulgation, and adoption of codes of conduct for scientists". The UK will chair these meetings, and as such professional organisations and others here have taken a particular interest in codes [37, 38, 39]. In 2002 the UN General Assembly and the Security Council endorsed the idea of establishing codes of conduct across areas of research relevant to the development of weapons of mass destruction. The suggestion for the BW area is now being taken up by the Inter Academy Panel and the International Centre for Genetic Engineering and Biotechnology [40]. The Policy Working Group on the UN and Terrorism had suggested that codes of conduct should:

> "Aim to prevent the involvement of defence scientists or technical experts in terrorist activities and restrict public access to knowledge and expertise on the development, production, stockpiling and use of weapons of mass destruction or related activities" [39, p. 14]

The U.S. NSABB has been tasked with developing "Professional codes of conduct for scientists and laboratory workers that can be adopted by professional organisations and institutions engaged in life science research" [41, p. 1].

Despite the widespread attention to developing a code or codes, at this stage it is not clear what they would entail in terms of their aims and audiences. Overall, professional scientific codes (whether they be called "code of conduct", "practice" or "ethics") vary greatly in terms their content and purposes. They extend from statements that lay out aspirational goals in the desire to raise awareness of issues or set ideal standards to more advisory or educational guidelines that provide suggestions for action to enforceable rules that specify do's and don'ts with enforcement mechanisms attached (e.g. steps regarding the physical control of lab materials) [42].[6] Initial statements by governments and others about the need for a "code of conduct" include reference to this range of types.

Certainly, the effectiveness of codes in general (not specific to bioweapon concerns) has been questioned by social scientists and ethicists. They are often open to numerous interpretations. Codes are rather inert forms of knowledge that are limited in their ability to deal with complex issues. They can be PR exercises that stave off other forms of regulation. Codes require enforcement mechanisms but to the extent that they are binding then the issue for evaluation is the merits of regulation in question [43, 44, 45, 46]. More positive analyses of codes in professional life have maintained that they can heighten awareness of issues, help individuals re-interpret the situations facing them, set expectations, clarify individual versus collective forms of responsibility, and influence behaviour for areas where standards have not formed [47, 48, 49, 50].

The purpose and content of such codes deserves considered attention. If the primary threat from bioweapons is conceived of in terms of terrorist attacks, then at first glance ethical codes would seem an incredibly ill-suited policy option. Yet given the concern with the dual-use potential of wide ranging areas of life science research undertaken by professionals in universities and industry, codes may have varied roles to play even with respect to this concern. While it is highly doubtful any would-be terrorists would be put off by the formation of codes, they could go some way towards forming a workable collective agreement or (minimally) encouraging the presentation of alternative points regarding how to address the problems, dilemmas and issues discussed in Section 2.

Several traits are likely to characterise useful code-related activities. First, much of the existing policy discussion speaks to the desired content of codes (read: outputs).

However, in keeping with the conclusions of studies of other professional codes, the process of devising and adopting a code needs concerted attention. Certainly outside of the U.S., substantial policy dialogue about the wisdom of research controls on knowledge and techniques is in its initial phase. Arguably beyond those areas of research working with select agents, attention to dual-use issues in the past has been minimal. Properly handled discussions on whether or what kind of codes would be appropriate can be one way of facilitating structured discussion. This should not be limited to practicing bioscientists as it has often been framed in policy documents to date. Rather it should include those who fund, represent, administer, commercialise, and regulate research in order to facilitate the sort of systematic and synergistic thinking required. Through this, the extent of the dual-use "problem" within the life sciences could be explored and debated.

Second, codes can do much more than merely keeping the conversation going about bioweapons. The adoption of codes could go beyond simply reiterating the existing provisions of national and international regulations and conventions (though this might prove useful in and of itself) by seeking to elaborate their meaning. There is much ambiguity in the interpretation of many provisions, much of it a product of the consensual character of international negotiations. Professionals in the life sciences can contribute to reaching a working agreement on their meaning. So just what the provisions of the BTWC mean in relation to the development of so-called non-lethal incapacitating agents or what counts as permissible biodefence activities are matters of disagreement today. Such matters are ones where those outside of traditional security networks can and should contribute to current debates [51]. That might be in specifying do's and don'ts or establishing expectations about the communication and transparency of such activities necessary to build international confidence in their legitimacy. Another clarifying function that code-related processes could do is to comment on who counts as the relevant community for make decisions about controls. Part of that might entail asking in what ways the "prevention against biowarfare and bioterrorism [is] too important to leave to the scientists and politicians" [52].

Third, in whatever they call for, codes should attempt to build on existing practices and controls in some way. To take the debate forward, codes could seek to make issues about biological weapons relevant to a wide range of activities currently not affected by biosecurity regulations so as to make the issues of dual-use more prevalent in the day-to-day practices of the life sciences. This need not require adopting burdensome forms of regulation, but instead building concerns about dual-use issues within the routines of research practice.[7]

Codes are unlikely to resolve debates about what needs doing and what is problematic about research. However, approached in the general manner outlined above and the subject of sufficient effort, then the process of adopting codes could play a role in fostering long term and focused discussion across the life sciences about many of the difficult issues raised in Section 2.

5. Conclusion

Today, wide ranging concerns are being expressed about the dual-use applications of the life sciences. The potential for a terrorist attack with advanced bioweapons capacities is a topic of much political and public concern. However low the probability of such an attack, it would likely be highly consequential. In a manner that is perhaps unprecedented, the life sciences in industrialised nations are being scrutinised for their ability to further nefarious ends. This Chapter has sought to challenge the aims and direction of many of the current controls being pursued for results and techniques. With regard to this issue, it has been argued that current discussions are unclear and arguably problematic in relation to vital questions regarding the problem posed by life science research, the ultimate goals of controls, the desirability of the circulation of dual-use knowledge, and the matter of who should decide what constitutes dangerous research. This conclusion challenges simple notions both that nothing and too much is happening. Rather, the picture presented is one where much uncertainty and disagreement exists regarding the way forward. Consideration has been given to the potential and pitfalls associated with current international efforts to devise a "code of conduct" for bioscientists. The Chapter has presented a basic outline of the desirable characteristic of the adaptation of codes to address many of the policy questions being discussed today.

Notes

1 The work undertaken for this paper was funded by the UK Economic and Social Research Council (ESRC) New Security Challenges Programme (RES-223-25-0053).

2 Though see publications such as [6]

3 Ciproflaxin is an antibiotic for anthrax exposure.

4 That is, so long as the necessary safety measures are undertaken.

5 Information about BW codes is available at http://www.ex.ac.uk/codesofconduct/

6 See www.codesofconduct.org for many written examples.

7 For instance, Pearson has suggested building on existing national and international health and safety regulations by establishing a code of practice requiring that current hazard assessments of scientific activities also consider their legality under obligations deriving from prohibition agreements [53].

References

[1] Brickley P. US universities resisting government attempts to control fundamental research. The Scientist; 7 January 2003. Available at: www.the-scientist.com

[2] Nixdorff K, Bender W. Ethics of university research, biotechnology and potential military spin-off. Minerva 2002; 40: 15-35.

[3] National Research Council (US) [NRC]. Biotechnology Research in an Age of Terrorism, Committee on Research Standards and Practice to Prevent the Destructive Application of Biotechnology. Washington, DC: National Academies Press; 2004.

[4] Tucker J. Preventing the misuse of pathogens. Arms Control Today 2003; June: 3-10.

[5] Block S. Facing the growing threat of biological weapons. 42nd Annual Meeting of the American Society for Cell Biology; 14 December 2002.

[6] Biosecurity and Bioterrorism. Istituto Diplomatico Mario Toscano, Villa Madama, Roma, September 18-19, 2000. See http://www.mi.infn.it/~landnet/Biosec/

[7] Leitenberg M. An assessment of the threat of the use of biological weapons or biological agents. In: Martellini M, editor. Biosecurity and Bioterrorism. Washington, DC: Landau Network Centro Volta; 2000. See http://www.mi.infn.it/~landnet/Biosec/

[8] Steinbruner J, Harris E, Gallagher N, Gunther S. Controlling Dangerous Pathogens: A Prototype Protective Oversight System; 5 February 2003. Available at www.cissm.umd.edu/documents/pathogensmonograph.pdf.

[9] Scientific Openness and National Security Workshop. The US National Academies and the Center for Strategic and International Studies; 9 January 2003; Washington DC.

[10] Shea D. Balancing Scientific Publication and National Security Concerns. Washington, D.C.: Congressional Research Service; 10 January 2003.

[11] Knezo G. "Sensitive but Unclassified" and Other Federal Security Controls on Scientific and Technical Information. Washington D.C.: Congressional Research Service; 2 April 2003.

[12] Journal Editors and Authors Group. Proceedings of the National Academy of Sciences 2003; 100 (4): 1464.

[13] Morse S. Bioterror R&D. Presented at Scientific Openness and National Security Workshop. National Academies and the Center for Strategic and International Studies; 9 January 2003; Washington, D.C.

[14] Relyea H. Silencing Science. Norwood, NJ: Ablex Publishing; 1994.

[15] Shouse B. US cell biologists urged to resist government limits on research. The Scientist; 16 December 2002.
[16] Sunshine Project Biosafety Bites Short Series. Available at: at http://www.sunshine-project.org/
[17] Cello C, Paul A, Wimmer E. Synthesis of poliovirus in the absence of a natural template. Presented at meeting on national security and research in the life sciences. National Academies and the Center for Strategic and International Studies; 9 January 2003; Washington, D.C.
[18] Rosengard A. Sensitive research studies – I. Presented at meeting on national security and research in the life sciences. National Academies and the Center for Strategic and International Studies; 9 January 2003; Washington, D.C.
[19] Shenk T. Sensitive Research. Presented at meeting on national security and research in the life sciences. National Academies and the Center for Strategic and International Studies; 9 January 2003; Washington, D.C.
[20] Cozzarelli N. PNAS policy on publication of sensitive material in the life sciences. Proceedings of the National Academy of Sciences 2003; 1000(4): 1463.
[21] Schuler A. Billions for biodefense. Biosecurity and Bioterrorism 2004; 2(2): 86-96.
[22] Leitenberg M. Biological Weapons in the Twentieth Century: a Review and Analysis. Available at http://www.fas.org/bwc/papers/21centurybw.htm.
[23] Turner J. Beyond Anthrax: Confronting the Future Biological Weapons Threat. Available at: http://www.house.gov/hsc/democrats/
[24] Smith B, Inglesby, T, O'Toole T. Biodefense R&D. Biosecurity and Bioterrorism 2003; 1(3): 193-202.
[25] Korch G. Leading edge of biodefense: the National Biodefense Analysis and Countermeasures Center. Department of Defense Pest Management Workshop; February 9-13 2004; Jacksonville, FL.
[26] Rappert B. Coding ethical behaviour: the challenges of biological weapons. Science and Engineering Ethics 2003; 9(4): 459-462.
[27] Müllbacher A, Logbis M. Creation of killer poxvirus could have been predicted. Journal of Virology 2001; September: 8353-5.
[28] Dennis C. The bugs of war. Nature 2001; 411; 17 May: 232-5.
[29] Gilbert N, Mulkay M. Opening Pandora's Box. Cambridge: Cambridge University Press; 1984.
[30] Collins H. The seven sexes. Sociology 1975; 9: 205-24.
[31] Petro J, Plasse T, McNulty J. Biotechnology: impact on biological warfare and biodefense. Bioterrorism and Biosecurity 2003; 1(3): 161-168.
[32] Poste G. The life sciences. Presented at meeting on national security and research in the life sciences. National Academies and the Center for Strategic and International Studies; 9 January 2003; Washington, D.C.
[33] Carlson R. The pace and proliferation of biological technologies. Biosecurity and Bioterrorism 2003; 1(3): 203-214.
[34] Couzin J. Bioterrorism: a call for restraint on biological data. Science 2002; 297; 2 August: 749-751.
[35] Block S. Facing the growing threat of biological weapons. 42nd Annual Meeting of the American Society for Cell Biology; 14 December 2002.
[36] Rappert B. Responsibility in the life sciences: assessing the role of professional codes. Biosecurity and Bioterrorism 2004; 2(3): 164-175.
[37] House of Commons Foreign Affairs Committee. The Biological Weapons Green Paper. HC 150; First report of session 2002-03; 11 December 2002.
Available at http://www.publications.parliament.uk/pa/cm200203/cmselect/cmfaff/150/150.pdf.
[38] Wellcome Trust. Wellcome Trust position statement on bioterrorism and biomedical research; 2003. Available at www.wellcome.ac.uk/en/1/awtvispolter.html.
[39] Royal Society, Paper on the Individual and Collective Roles Scientists can Play in Strengthening International Treaties, Policy Document 05/04, April 2004.
[40] United Nations General Assembly Security Council [UN]. Report of the Policy Working Group on the United Nations and Terrorism. New York: UN; 6 August 2002. Available at: http://www.un.dk/doc/A.57.0273_S.2002.875.pdf
[41] National Science Advisory Board for Biosecurity. 'NSABB Charter' Available at: http://www.biosecurityboard.gov/
[42] Soskolne C, Sieswerda L. Implementing ethics in the professions. Science and Engineering Ethics 2003; 9: 181-190.
[43] Iverson M, Frankel M, Siage S. Scientific societies and research integrity. Science and Engineering Ethics 2003; 9: 141-158.
[44] Doig A, Wilson J. The effectiveness of codes of conduct. Journal of Business Ethics 1998; 7(3): 140-149.
[45] Higgs-Kleyn N, Kapelianis D. The role of professional codes in regulating ethical conduct. Journal of Business Ethics 1999; 19: 363-374.

[46] Shrader-Frechette K. Ethics of Scientific Research. Lanham: Rowan & Littlefield; 1994.
[47] Davis M. Thinking Like an Engineer. Oxford: Oxford University Press; 1998.
[48] Meselson M. Averting the exploitation of biotechnology. FAS Public Interest Report 2000; 53: 5.
[49] Unger S. Code of engineering ethics. In: Johnson D (ed.) Ethical Issues in Engineering. Upper Saddle River, NJ: Prentice Hall; 1991: 105-130.
[50] Reiser S, Bulger, R. The social responsibilities of biological scientists. Science and Engineering Ethics 1997; 3(2): 137-143.
[51] Rappert B. Non-Lethal Weapons as Legitimizing Forces? Technology, Politics and the Management of Conflict. London: Frank Cass; 2003.
[52] Pax Christi International. Statement: Pax Christi International calls for ethical approach to biological weapons. Ref.: SD.08.E.04; Brussels; June 2004.
[53] Pearson G. A Code of Conduct for the Life Sciences: A Practical Approach. Bradford Briefing Series No 15. Bradford: Department of Peace Studies, University of Bradford; 2003. Available at: http://www.brad.ac.uk/acad/sbtwc/briefing/BP_15_2ndseries.pdf

Subject Index

Science and Technology Policies for the Anti-Terrorism Era
Andrew D. James (Ed.)
IOS Press, 2006

Author Index